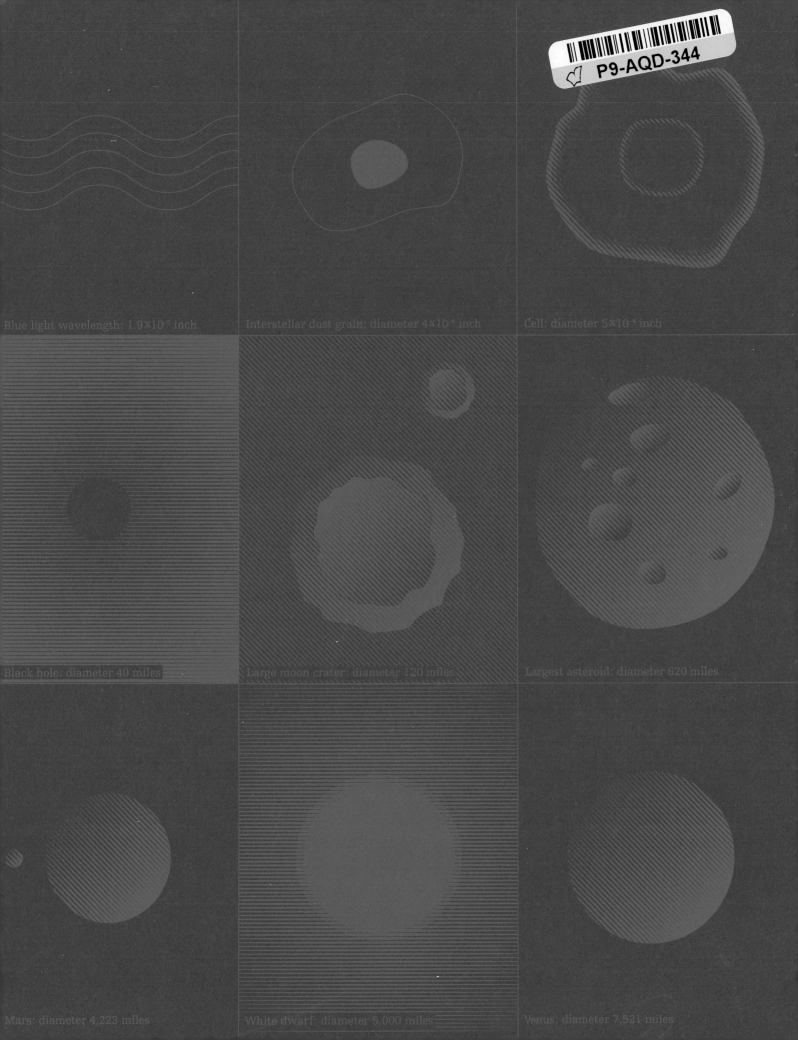

Blue light wavelength: 1.9×10⁻⁵ inch

Interstellar dust grain: diameter 4×10⁻⁶ inch

Cell: diameter 5×10⁻⁴ inch

Black hole: diameter 40 miles

Large moon crater: diameter 120 miles

Largest asteroid: diameter 620 miles

Mars: diameter 4,223 miles

White dwarf: diameter 5,000 miles

Venus: diameter 7,521 miles

STARBOUND

The glowing exhaust trails of engines fueled by matter-antimatter annihilation illuminate the inner surfaces of three ring-shaped interstellar arks crossing the ocean of.darkness between suns. As envisioned by an aerospace engineer, each vessel, rotating enough to create the equivalent of Earth's gravity, carries some 1,600 space pioneers in four pairs of habitation towers spaced at intervals along the ship's 29-mile-long perimeter.

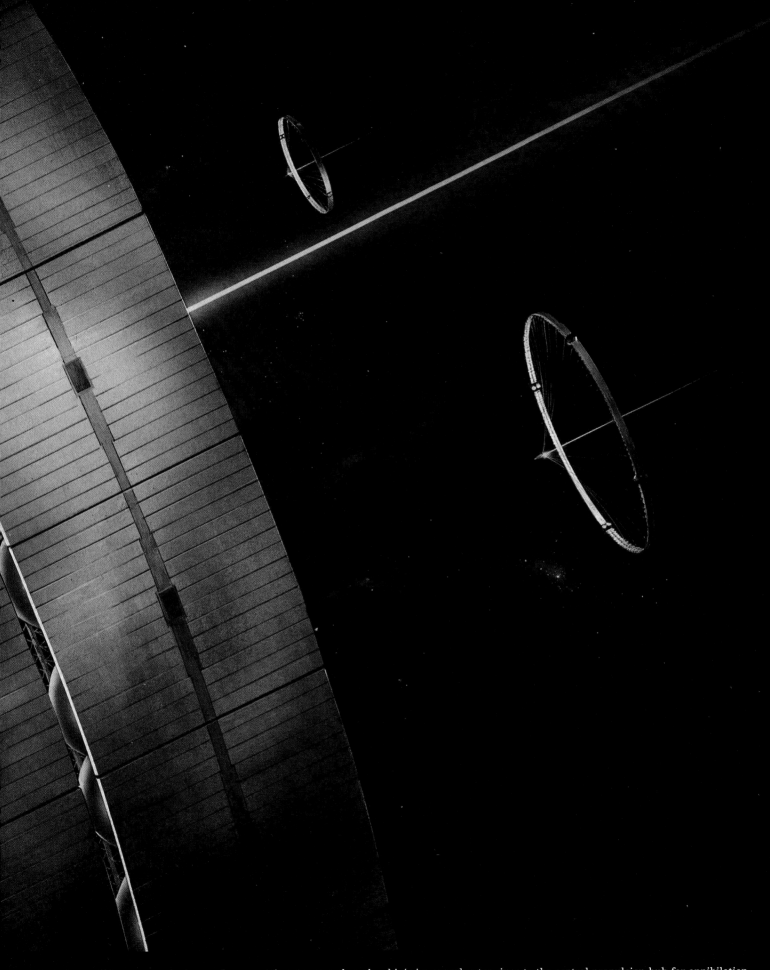

Spokelike fuel lines convey hydrogen from containers mounted on the ship's inner and outer rings to the central propulsion hub for annihilation with stores of antihydrogen ice. Traveling at two percent of the speed of light, or 3,720 miles per second—the theoretical velocity of these ships—the fleet would take several generations to journey to another solar system.

TIME
LIFE
BOOKS

This volume is one of a series that examines the universe in all its aspects, from its beginnings in the Big Bang to the promise of space exploration.

The Cover: Separated from a nuclear reactor in the radiation cone by a two-mile-long structure containing fuel-efficient ion engines, a rotating carousel a thousand feet in diameter creates artificial gravity for travelers on an interstellar mission.

VOYAGE THROUGH THE UNIVERSE

STARBOUND

BY THE EDITORS OF TIME-LIFE BOOKS
ALEXANDRIA, VIRGINIA

CONTENTS

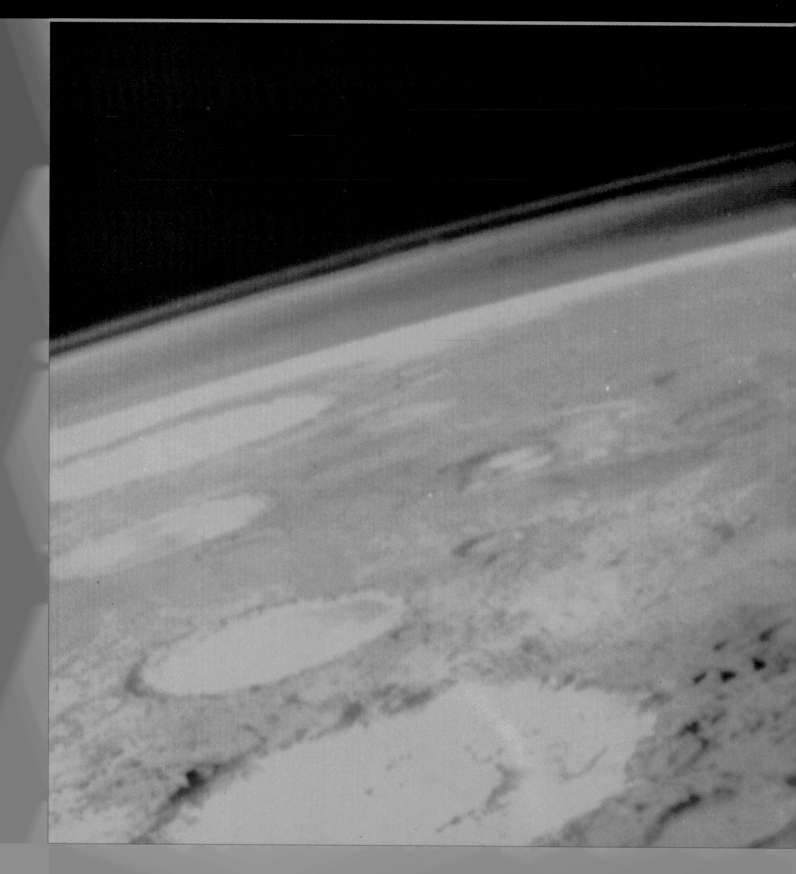

mile-wide Argyre basin *(below, left)*, the tenuous carbon dioxide atmosphere of Mars is scarcely more than a wispy haze on the horizon. But the Red Planet is a

journey toward the stars, as explorers from Earth learn to navigate within the Solar System—readying their ships and themselves for the next great step.

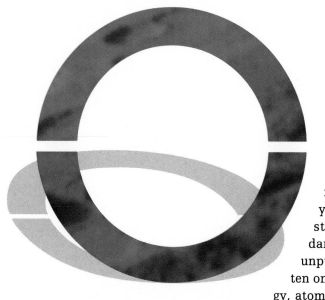

ne day in January 1918, a young American physics instructor named Robert Goddard gave a friend a collection of unpublished papers he had written on such subjects as solar energy, atomic power, and the avoidance of meteors by travelers in outer space. Goddard, thirty-five years old at the time, had long been fascinated by futuristic ideas. Growing up in the Boston suburb of Roxbury, Massachusetts, he had devoured Jules Verne's novel *From the Earth to the Moon.* Later, after his family had moved to nearby Worcester, he had listened rapt as the astronomer Percival Lowell lectured to a local audience on his theory that Mars was populated by highly intelligent beings.

But Goddard was more than just imaginative. He possessed practical genius. Before he was out of college, he began experimenting with rockets, and in 1914, he was issued two patents—one for a multistage rocket design and the other for rocket fuel systems that used both liquid and solid propellants. During the decades ahead, he would prove himself a peerless inventor in the field of rocketry, earning more than 80 patents during his lifetime; 131 were granted posthumously. Now, as he prepared to go to California on a wartime assignment for the government, he handed his papers to his friend for safekeeping. Goddard put the manuscript in an ordinary envelope, which he then placed inside another one that he labeled "Special Formulae for Silvering Mirrors." On the inner envelope he wrote, "The notes should be read thoroughly only by an optimist."

The injunction seemed especially apt for the musings he had titled "The Last Migration." Here, among what he would later refer to as his "most extreme speculations," were suggestions for how human beings could make the seemingly impossible journey to vastly distant stars. People could be put in what he described as "a state of granular protoplasm" that would allow their bodies to withstand "low temperatures for long intervals," just as seeds hibernate over winter. A pilot in this deep sleep could be awakened every ten thousand to a million years by a radium alarm clock activated by an increase in gas pressure as the radium decayed. Alternatively, generation upon generation of travelers might simply live and die aboard a starship, although, Goddard noted, "the characteristics and natures of the passengers might change" as millennia passed.

Goddard did not hazard a guess as to when this extraordinary vision of interstellar travel might become a reality, but his paper implied a time far in the future. Today, in part because of the sturdy foundation he built for the science of rocketry, the notion of journeying out along a series of stepping-stones to the stars is a matter of serious scientific discussion. Some people speak of it as a new and infinitely more momentous version of Manifest Destiny, the phrase applied to America's westward expansion in the nine-teenth century. Others cite the eloquent words of Konstantin Tsiolkovsky, a Russian schoolteacher and mathematician who envisioned orbiting space colonies and inspired a generation of scientists with his book *Beyond the Planet Earth,* published in 1920. Said Tsiolkovsky, "Earth is the cradle of mankind, but one does not live in the cradle forever." Freeman Dyson of the Institute for Advanced Study in Princeton, New Jersey, an eminent theoretical physicist and dreamer about the cosmos, has written of "the spreading out of life in all its multifarious forms from its confinement on the surface of our small planet to the freedom of a boundless universe."

AN EXPANSIONARY SPECIES

If the testing of these dreams lies out in the far future, the impulse that drives them can be traced to the deep past—to a time when humanness, or at least the first glimmerings of humanness, revealed itself in the long and tangled drama of life's evolution on Earth. Sometime between four and five million years ago, a new creature appeared on the open savannas of Africa. This hominid, which would be given the name *Australopithecus,* or "southern ape," was destined to be the forebear of the most expansionary of all earthly species. Descended from a tree-dwelling line of primates, australopithecines were small-brained but resourceful enough to use elementary tools of stone and wood. They lived off roots, tubers, insects, and any meat they could scavenge on their grassy habitat.

By about 1.6 million years ago, the little hominids had given rise to a species called *Homo erectus.* The members of this more advanced evolutionary line had a brain size that overlapped the lower end of the range for modern human intelligence. They made sharp-edged stone tools and their primary source of food was big game such as bison. As they traveled in small groups across the tree-dotted grasslands in search of their prey, they steadily expanded their range. This wandering took them far from their ancestral African home into the warm savannas that then stretched hospitably across southern Asia and into southern Europe.

The roving bands penetrated into radically different environments as well. By about 100,000 years ago, humans had begun to move not only into the dense rain forests of central Africa, but also into regions where water was scarce, and into the treeless tundras of Europe and Russia where big game abounded —great herds of bison, reindeer, horses, mammoths, mastodons, and other grazing animals. The hominids' great adaptive advantages were intelligence and language, which enabled them to learn and to transmit knowledge to descendants in a systematic way no previous species had ever done. Colder lands were conquered by a series of inventions: the controlled use of fire, the fashioning of protective clothing from animal skins, the building of shelter from wood, hides, and animal bones.

At the same time, the tribes were gaining new territorial opportunities. With the coming of the last ice age, 79,000 years ago, great quantities of water were locked up in vast ice sheets, causing the level of the oceans to drop by 300 feet or more and thereby increasing the planet's land area. Indonesia, once an unreachable scattering of islands, became an extension of Asia. Perhaps 50,000 years ago, humans dwelling along the coasts of this reclaimed region built simple vessels—rafts, most likely—and made their way from island to island across the Timor Straits to the continent of Australia, then a landmass of more than three million square miles.

As one of today's visionaries on the subject of interstellar travel, astrophysicist Freeman Dyson *(far right)* joins the ranks of Russian mathematician Konstantin Tsiolkovsky *(below)* and rocket scientist Robert Goddard. At the turn of the century, Tsiolkovsky conceived a liquid-fuel multistage space rocket. Two decades later, Goddard refined the theory of staging and began testing a series of experimental liquid-fuel rockets, one of which reached an altitude of 9,000 feet. Dyson, a maverick futurist, has proposed several schemes using nuclear bombs to power expeditions to the stars.

The fall in sea level also created a thousand-mile-long land bridge between Siberia and Alaska. About 12,000 to 14,000 years ago (the date is far from certain), bands of Siberian hunters followed the herds of grazing animals across the tundra bridge and eventually found a route south, where boundless riches lured them ever farther. By 9000 BC, the migrants to the New World had reached the tip of South America; those who had stopped at intermediate points en route swiftly mastered the demands of life in the mountains, deserts, forests, and grasslands of the Americas.

But the epic of human expansion was far from over. One of the most remarkable chapters got under way about 5,000 years ago, when the coastal peoples of southeastern Asia—the ancestors of today's Polynesians—began pressing the bounds of their range in seagoing craft. From the Indonesian and Philippine archipelagos, they traveled along the north shore of New Guinea

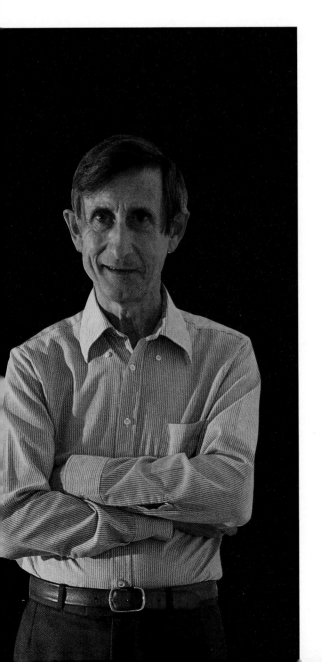

and settled some small offshore islands. Around 1500 BC, and within just a few generations, they spread 1,800 miles across the Pacific to the uninhabited islands of Fiji, Tonga, and Samoa. For their blue-water journeys, they developed extraordinary navigational skills, determining latitude by stars overhead, memorizing other stars for fore-and-aft bearings, detecting the presence of land by clouds hanging stationary in the sky or the patterns of bird flight or the refractive effect islands had on the regular Pacific swell. Their vessels of migration by now were huge wooden canoes, double-hulled or fitted with outriggers, and laden with food, plants that would be cultivated, and pigs, chickens, and dogs.

Most of the island-hopping required the Pacific pilgrims to cross no more than a hundred or so miles of open water, but sometimes they ventured thousands of miles into the unknown before encountering land—if they encountered it at all. In time, the oceanic realm of the Polynesians encompassed a huge triangular area, the corners marked by Hawaii, Easter Island, and New Zealand, that was almost the size of Europe and Asia combined. Virtually all of it was ocean: 995 parts in a thousand, according to one estimate.

THE OCEAN OF SPACE

However daunting that ratio of water to land, it pales in comparison to what the human species faces in the vast ocean of space. All the solid bodies of the Solar System—the planets, moons, and asteroids—together account for no more than an estimated 3×10^{-15} (that is, 3 preceded by 14 zeros and a decimal point) percent of the spherical volume occupied by the solar family. Yet these are crowded conditions compared to the general galactic neighborhood. The nearest stellar landfall—the three-star system known as Alpha Centauri—lies 4.3

light-years, or 25 trillion miles, from Earth. The next-closest sun, Barnard's Star, is 5.9 light-years away.

To cross this cosmic ocean, even to a body as near as the Moon, is an enterprise fraught with difficulties. The continents and islands on planet Earth, however vast may be the expanses of water between them, at least stay in one place for the duration of a journey. Planets, moons, and stars, by contrast, move and interact in a ceaseless gravitational dance that requires spacecraft to follow curving, looping trajectories painstakingly worked out according to Newton's laws of motion.

And the hazards go beyond the navigational. Travelers are vulnerable to lethal high-energy particles and radiation that flood space when magnetic

Celestial Chronometers

Since prehistoric times, humans have looked to the heavens to mark the natural rhythm of earthly events—the ceaseless turn of day and night, the march of the seasons. Attempts at calendar keeping may date back to some 37,000 years ago, when, evidence suggests, ice age hunter-gatherers first recorded the phases of the Moon. Through the eons, efforts to chronicle celestial motions grew increasingly complex as various cultures erected structures whose form and function were patterned on the stars themselves.

6500 BC The sequence of notches carved on this ice age antler tool may correspond to lunar phases. The broad end has been fashioned to resemble the head of an animal.

2800 BC The newly risen midsummer Sun gleams above the Heel Stone *(center boulder)* of England's Stonehenge, whose many lunar and solar alignments are subject to various scientific interpretations.

storms rage on the Sun; and rocky debris from asteroids or comets, zipping along at tens of thousands of miles per hour, can inflict serious, if not fatal, damage to the hull of a spacecraft. Even mundane concerns become highly problematic in space—the difficulties of communicating across the void, for instance, and the debilitating effects of living for long periods without the normal tug of Earth's gravity.

Yet the urge and the need to reach out for new territory are as strong as they have ever been in the genus *Homo:* Space, with all its inherent risks and dangers, is the last frontier. In 1961, while the world held its collective breath below, Yuri Gagarin of the Soviet Union became the first human being to brave that airless realm when he completed a single orbit of Earth in a craft scarcely large enough for one person. Within three decades, American astronauts had repeatedly visited the Moon, and Soviet cosmonauts had spent as much as a year at a time living in relatively spacious quarters aboard orbiting space stations. Today, the scientists and engineers of several nations are plotting next steps, examining their options for establishing bases on the Moon and Mars, considering the prospects for piloted journeys to asteroids or to the satellites of the outer planets, and studying the feasibility—the inevitability, some say—of departing the Solar System altogether and sailing out across the cosmos to planets circling other stars. Their deliberations range from detailed engineering studies involving known technologies, such as the construction

2500–1400 BC An Egyptian tablet depicts the pharaoh *(left)* and the goddess Seshat laying out a temple by aligning a cord with the Big Dipper, symbolized by the seven-pointed star above Seshat.

362–343 BC A chronicle of the motions of Mars and Mercury appears in cuneiform script in this clay Babylonian astronomical diary, one of thousands that record planetary and lunar events.

of a space station in low Earth orbit that will serve as a staging ground for subsequent journeys, to projects far more speculative, such as life-support systems for interstellar trips lasting hundreds of years.

GETTING THERE

The most fundamental challenge is propulsion—the means of getting there, whether "there" be the Moon (a mere three days away) or Mars (months away) or, at the farthest edge of possibility, other galaxies (millions of light-years distant). Chemical rockets, the standard vehicles for all space travel up to now, will play the key role for the next several decades if current plans are borne out, but they may someday burn fuel mined on the Moon or on asteroids. Many space scientists are urging the development of propulsive systems that rely on electrical effects—creating thrust by electrostatic acceleration of ions or by electromagnetic acceleration of plasma; the electricity itself would be generated by lasers or by nuclear-fission power plants. Some astronautical engineers, seizing on technology that already exists, are designing solar sails, huge expanses of aluminized plastic that will unfurl in space and be propelled by the feather-light but relentless pressure of photons radiating from the Sun *(pages 18-23).* Others are sketching ideas for engines that will work by nuclear fusion, the atom-melding process that lights the Sun. Still others foresee supremely efficient and powerful engines that will exploit the mutual annihilation of matter and antimatter.

Underlying all the questions of how is the issue of why. What could possibly justify the risk of life or the expenditure of billions or trillions of dollars to carry human beings beyond their home planet? Knowledge is one obvious

300 BC At Namoratunga II, an observatory in northwest Kenya, basalt pillars point to the rising positions of stars still used by the Eastern Cushite people in reckoning their calendar.

AD 1000 Triangles of light cast at sunset on the equinox, when day and night are equal in length, form an undulating serpent on a stairway of the Castillo, a pyramid at Chichén Itzá in the Yucatán.

prize. A bounty of astronomical information could be gleaned from an observatory on the Moon, for example, and clues found on other planets could help unravel the early history of the Solar System. Scientists might also be able to resolve such long-debated issues as the existence of life on Mars; the Viking probes to the Red Planet in the 1970s conducted a worthy investigation, but robotic explorers cannot equal the eyes and inquiring minds of on-the-scene humans. Other potential rewards of leaving Earth's haven are of a more material nature. Valuable substances, such as carbon, nickel, gold, platinum, phosphorus, and iron, may someday be extracted from asteroids or distant moons.

Finally, however, the settlement of other worlds has a profound strategic dimension for the human species: Dispersion would enhance the chances of surviving terrestrial disaster. Degradation of Earth's biosphere by pollutants may someday reach fatal levels. A terrible war could poison the planet with highly radioactive fallout. Or the blind forces of nature could deliver a devastating blow in the form of an asteroid impact—a trauma that has happened repeatedly in the past and that is virtually certain to happen again.

Thousands of asteroids periodically cross Earth's orbit and are candidates for collision, but astronomers have plotted the paths of only a few dozen of them. On March 23, 1989, without any warning, an object about a third of a mile in diameter and traveling at tens of thousands of miles per hour passed within half a million miles of Earth—a very near miss by astronomical standards. If it had struck the planet's land surface, it would have exploded with a force equal to a million tons of TNT and gouged a crater more than four miles

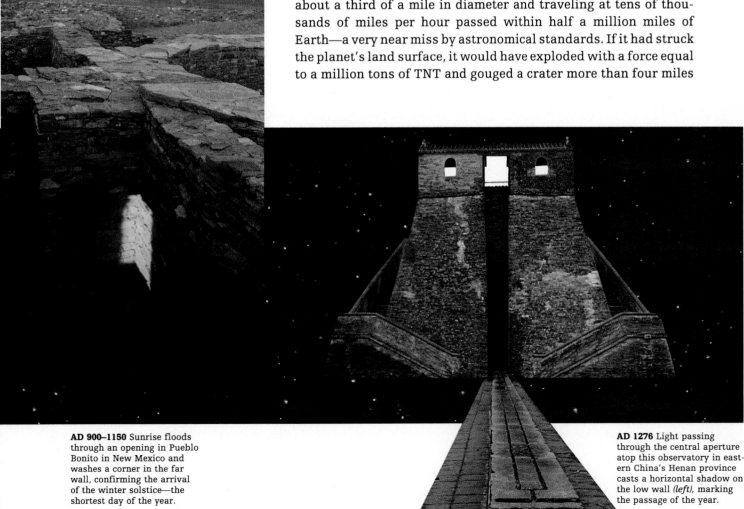

AD 900–1150 Sunrise floods through an opening in Pueblo Bonito in New Mexico and washes a corner in the far wall, confirming the arrival of the winter solstice—the shortest day of the year.

AD 1276 Light passing through the central aperture atop this observatory in eastern China's Henan province casts a horizontal shadow on the low wall *(left)*, marking the passage of the year.

Sun-Powered Flight

Just as the fifteenth century's Great Age of Exploration relied on the pressure of the wind on cloth sails to propel fleets of ships across vast terrestrial oceans, space-age explorers may someday harness light itself for journeys across the airless void between planets—or even between planets circling a distant star. Although photons, the packets of electromagnetic energy that radiate from the Sun and all stars, are essentially massless, they can nonetheless apply a steady, windlike pressure in the near vacuum of space.

The key to exploiting sunlight's feathery touch for solar sailing, as this method of travel is known, is material that is highly reflective, so that incoming photons will deliver thrust by bouncing off the sail surface *(pages 22-23)*. The greater the ratio of sail surface area to the craft's total mass and the more photons it reflects, the greater the thrust.

However vast, though, the sail must also be rigid enough to make the craft steerable, a consideration that affects its final shape. The disk-shaped sail shown here is only one of many designs on drawing boards around the world, and the schemes vary with differing techniques for deployment and manipulation *(pages 20-21)*. Initially, the sails are meant to carry communications equipment and scientific experiments. Later, they would ferry supplies between Earth and Mars or other planets, and—in time—travelers may drift on sunlight from one interplanetary destination to another.

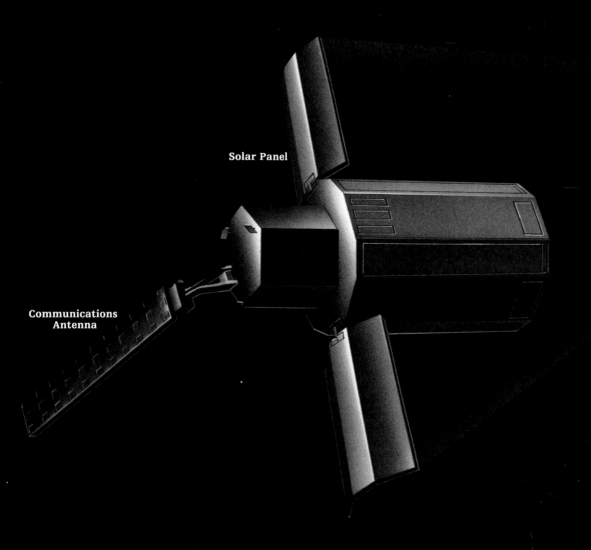

Solar Panel

Communications
Antenna

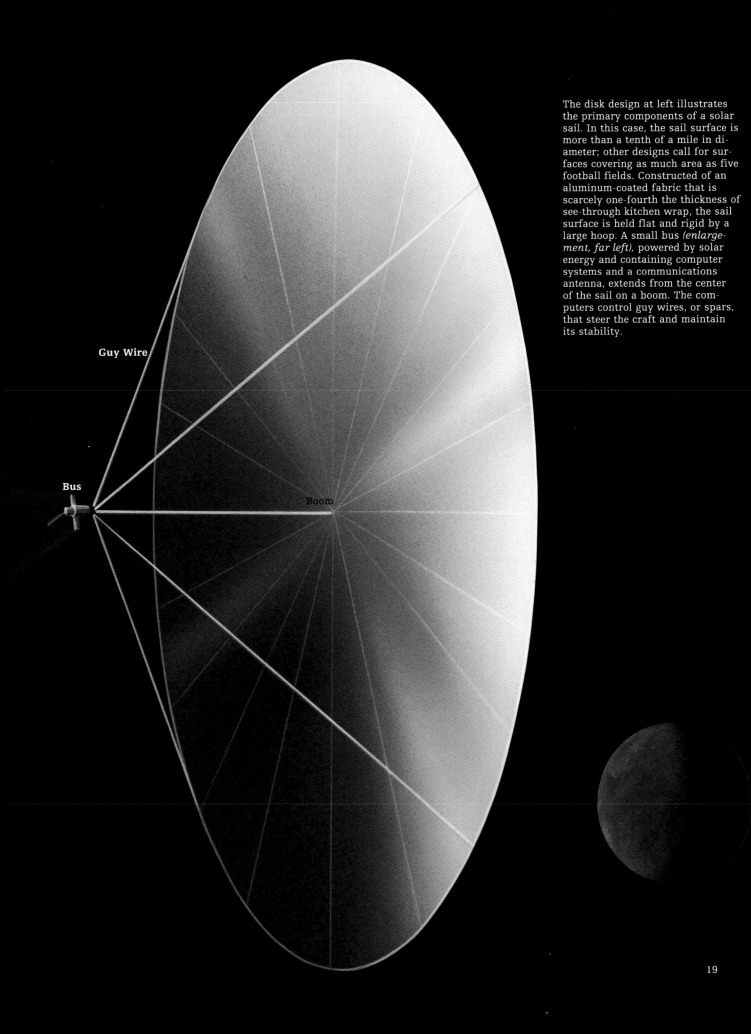

Guy Wire

Bus

Boom

The disk design at left illustrates the primary components of a solar sail. In this case, the sail surface is more than a tenth of a mile in diameter; other designs call for surfaces covering as much area as five football fields. Constructed of an aluminum-coated fabric that is scarcely one-fourth the thickness of see-through kitchen wrap, the sail surface is held flat and rigid by a large hoop. A small bus *(enlargement, far left),* powered by solar energy and containing computer systems and a communications antenna, extends from the center of the sail on a boom. The computers control guy wires, or spars, that steer the craft and maintain its stability.

A GALLERY OF SAIL DESIGNS

So weak is sunlight as a propulsive force that today's solar sail designers are faced with a dilemma: On the one hand, they want their craft to deploy as much sail surface as possible to capture a maximum of photons. However, to subject a flight-ready solar sail—with its five to six acres of fragile reflective material—to the violence of a terrestrial launch would be to court certain disaster. Because on-orbit construction is not yet feasible, sail engineers across the globe have come up with a number of ideas for packaging the sail and its payload to survive a rocket liftoff and then to deploy

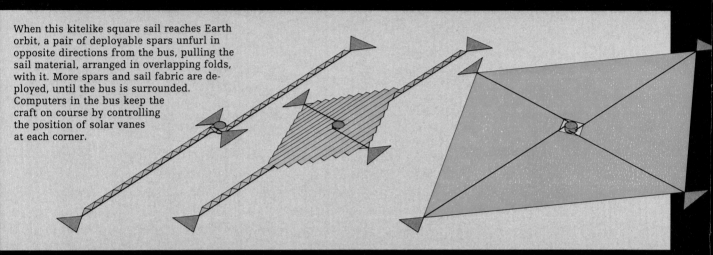

When this kitelike square sail reaches Earth orbit, a pair of deployable spars unfurl in opposite directions from the bus, pulling the sail material, arranged in overlapping folds, with it. More spars and sail fabric are deployed, until the bus is surrounded. Computers in the bus keep the craft on course by controlling the position of solar vanes at each corner.

This disk is sent into orbit with its sail folded along reinforced seams and wrapped, umbrella-like, around the bus. At deployment, flexible plastic spokes, embedded in the sail along some of the folds, spring open, fanning out the sail to its full dimensions—the area covered by three football fields. A motor in the bus positions the spokes to create several different shapes (the configuration known as the turbine is shown here), which allow the sail to be set and maintained at appropriate angles to the light.

automatically once safely in orbit at an altitude of about a thousand miles.

A critical consideration in bundling the aluminized sail material is how—or whether—to fold it. A hard crease in the sail's metallic surface would not only diminish its reflective capacity but also weaken the fabric, leaving it prone to tears. In addition, because the sail must be able to assume full size and shape by a foolproof mechanism once it is in orbit, any rigidizing structure, such as a hoop for a disk sail, must be compacted and ready to open as well.

Each design in the array shown here addresses such issues as ease of manufacturing, packaging, and deployment, as well as responsiveness of flight controls and speed of acceleration. All are theoretically capable of sailing an interplanetary course.

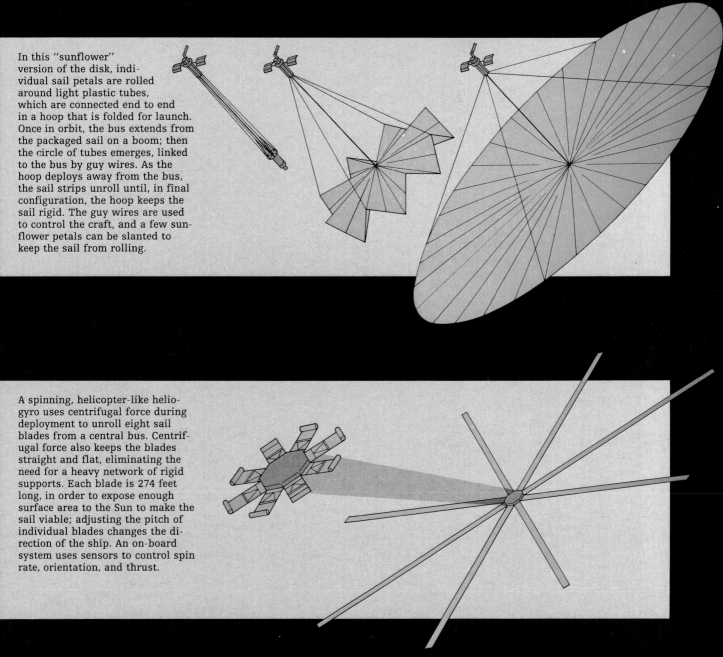

In this "sunflower" version of the disk, individual sail petals are rolled around light plastic tubes, which are connected end to end in a hoop that is folded for launch. Once in orbit, the bus extends from the packaged sail on a boom; then the circle of tubes emerges, linked to the bus by guy wires. As the hoop deploys away from the bus, the sail strips unroll until, in final configuration, the hoop keeps the sail rigid. The guy wires are used to control the craft, and a few sunflower petals can be slanted to keep the sail from rolling.

A spinning, helicopter-like heliogyro uses centrifugal force during deployment to unroll eight sail blades from a central bus. Centrifugal force also keeps the blades straight and flat, eliminating the need for a heavy network of rigid supports. Each blade is 274 feet long, in order to expose enough surface area to the Sun to make the sail viable; adjusting the pitch of individual blades changes the direction of the ship. An on-board system uses sensors to control spin rate, orientation, and thrust.

WINDING OUT OF THE SOLAR SYSTEM

Unlike conventional spacecraft, which require powerful rockets to boost them to escape velocity, a solar sail ship can be launched by a less-powerful vehicle to the necessary speed gradually over several weeks. As shown below in the scenario of a disk sail deployed in Earth orbit, the craft accelerates only during the portion of its orbit when it is heading away from the Sun. (The diagrams at bottom represent a view from above of the steps in the large illustration.)

On the outbound leg (1), the sail's mirrored surface

To build momentum to escape velocity, a solar sail ship must present the optimal amount of surface area to the Sun's radiation for a given position in the craft's orbit. As photons hit and bounce off *(yellow arrow)*, the resulting direction of thrust *(red arrow)* is approximately perpendicular to the sail surface; the more reflective the sail, the

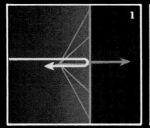

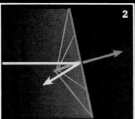

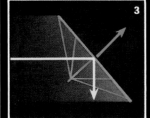

fully faces the solar radiation *(large arrow)*, reflecting the greatest possible number of photons to get maximum acceleration. Just before the turnaround (2), guy wires pull the bus off center, and the sail tilts so that photons strike at an angle, causing the craft to begin pivoting. Although the bus soon shifts back to center (3), the craft continues to roll, and by the time it is heading sunward again, it has turned over completely

(4). In this mode, the sail presents the narrowest profile (5, 6) as it travels against the outpouring of solar photons; it thus minimizes the loss of its store of momentum. Rounding Earth (7), the spars again pull the sail into position for another circuit (8, 9). In the course of a hundred or so orbits, the ship will spiral farther and farther away from Earth, until it finally musters the critical momentum to escape altogether.

closer to perpendicular the angle of thrust. For example, when photons hit the sail straight on (1), they reflect straight back, and the sail is pushed in the same direction as the radiation. Thus, by varying its angle of motion, the craft can accelerate on different portions of the outbound leg (2, 3, 8, 9) and conserve its momentum on the inward stretch (4, 5, 6, 7) when it is, in effect, heading upwind.

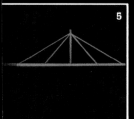

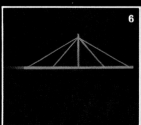

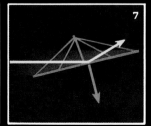

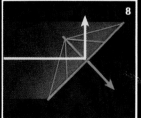

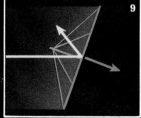

in diameter. Such an impact would have kicked up enough dust to reduce global temperatures precipitously and thereby cause grievous damage to crops, with repercussions that would have been felt throughout civilization.

The blow would have been the merest tap compared to some past collisions. Much evidence indicates that the Earth was struck 65 million years ago by a meteorite roughly six miles in diameter and possessing a mass of a trillion tons. Hurtling through the atmosphere at 55,000 miles per hour, it hit with the force of five billion atom bombs, blasting a crater ninety miles wide and darkening the planet with a pall of dust that lingered for as long as twelve months. Raging fires burned much of the world's vegetation, and chemical reactions caused the atmosphere itself to burn and then drench the Earth with acid rain. Many species of land and sea animals disappeared, including several dozen dinosaur species; although more adaptive species managed to survive the impact, the vast majority of their members died in the holocaust and the long winter that followed. A similar mass extinction at the end of the Permian period about 230 million years ago seems to correlate with a strike that might have been even worse.

Scientists have calculated that "civilization-threatening" impacts—the sort of damage that would have been associated with the asteroid that passed in 1989—can be expected about once every hundred thousand years. The evidence suggests that more calamitous blows of the kind that wiped out the dinosaurs come at intervals of tens or hundreds of millions of years. But all such events are essentially unpredictable. The Earth could be bludgeoned to near sterility tomorrow.

There is reason, then, to look outward, to widen humanity's sphere in the age-old frontier way. And even if the payoff is debatable or speculative, the human spirit tends to be incorrigibly venturesome. If a road is open or can be made open, humans will travel it.

RETURN TO THE MOON
Wherever it leads, the journey outward will have to begin with baby steps. A 1988 National Aeronautics and Space Administration report titled "Beyond Earth's Boundaries" speaks of the twenty-first century as "a time when humanity will have broken free of the physical and psychological bonds of planet Earth to live and work for extended periods on nearby bodies in our Solar System." But before that vision can possibly be realized, scientists, engineers, and world leaders must prepare the way.

Very likely, the first arena for intensive action will be the Moon. Since 1972, when the last Apollo mission lifted off from the lunar surface for the trip home, neither the United States nor the Soviet Union has paid much attention to Earth's nearest neighbor. The great expense of the Apollo project—$24 billion—was a daunting argument against further lunar ventures for the moment, and both nations limited their piloted space activities to low Earth orbit (LEO). The Americans concentrated on the development of a reusable vehicle, the space shuttle, while the Soviets devoted their efforts to building a space station, *Mir,* and studying the physiological effects of long-term weightlessness. Neither the shuttle nor *Mir* constituted an end in itself, however, and planners continued to dream of a day when humans would again set foot on the Moon, this time to stay.

Then, in the late 1980s, as the two superpowers broke through to a new diplomatic understanding, calls to action on the space frontier began to be sounded at the highest levels. In a speech delivered in July 1989, on the twentieth anniversary of the first Moon landing, President George Bush promised a national commitment to "a sustained program of manned exploration of the solar system and, yes, the permanent settlement of space." Just over a year earlier, Soviet President Mikhail Gorbachev had adopted a more global perspective, suggesting that his country and the United States work together on sending a crew to Mars.

By 1990, NASA had settled on a coordinated space strategy for the relatively near future. First, in the latter part of the decade, there would be a permanently staffed space station in low Earth orbit; then, just after the turn of the century, a return to the Moon and the establishment of a lunar base. The space station, orbiting 250 miles above the Earth, will serve in a variety of ways: as a vantage point from which scientists can study their home planet; as a gravity-free laboratory for experimentation and materials research; as a way station for the refueling and repair of satellites and a variety of space

After a seventy-five-hour mission in December 1972, the lunar module *Challenger,* standing sentinel here over the Moon's mountainous Taurus-Littrow terrain, carried two *Apollo 17* astronauts back into lunar orbit for the subsequent return to Earth. It was the last time humans set foot on the Moon's surface. After a long hiatus, NASA scientists foresee a renewed role for Earth's satellite in the conquest of space, and have devised an ambitious program calling for a permanent Moon base by the early twenty-first century.

SPACE WORKHORSES

To make working in space practical, engineers are designing unpiloted heavy lift launch vehicles (HLLVs), similar to the one shown at left. This version would hoist as much as 200,000 pounds of building materials, fuel, and equipment to construction sites in low Earth orbit, 220 miles above the surface. To take advantage of existing space shuttle technology and equipment, the vehicle is powered by shuttle-type solid rocket boosters that would be jettisoned after launch.

The maintenance of space observatories and other satellites could be handled by remotely piloted space tugs called orbital maneuvering vehicles (OMVs). Fitted with video cameras and grappling devices, the fifteen-foot-wide craft *(right)* would be used by operators at the space station or in the space shuttle to capture and repair ailing craft, or, as shown here, to go after satellites whose orbits have decayed and either reboost them to higher altitudes or tow them in for repair.

Long-range orbital transfer vehicles (OTVs) would carry personnel, equipment, and satellites back and forth between low Earth orbit and geosynchronous orbit, 22,300 miles high, with possible excursions to the Moon as well. On piloted missions, crew members would ride in a capsule mounted atop the saucer-shaped vehicle (below). At other times, the crew capsule might be removed to allow large satellites to be handled telerobotically.

vehicles; and eventually as a construction facility for the huge ships that will be used in the program of extraterrestrial settlement, including the activities on the Moon. Plans for the lunar base call for astronauts to arrive sometime in the first quarter of the twenty-first century, to be followed by several years of construction. The total enterprise would require expenditures of at least $150 billion over two decades—about as much as the Apollo program cost, in constant dollars.

Already, a new phase of lunar reconnaissance has been undertaken, although not by the United States. In early 1990, Japan became the third nation to launch a mission to the Moon. Boosted skyward from the Kagoshima Space Center on Kyushu, the southernmost island of Japan, on January 24, the *Hiten ("Sky Flight")* probe, which carried a micrometeoroid detector built in West Germany, was basically a test vehicle designed to try out techniques for reaching lunar orbit.

Japan's Institute of Space and Aeronautical Sciences (ISAS) has plans for more ambitious efforts, including a "penetrator" mission that would fire three javelin-like projectiles into the lunar surface. The penetrators would carry seismometers that will be used to measure the Moon's internal activity and to record the shock waves of meteorite strikes and moonquakes, and another instrument to measure heat flow from the lunar interior. Another ISAS mission is designed to put a satellite carrying multiwavelength spectrometers into polar orbit around the Moon in order to map its mineralogical composition. NASA is planning to deploy similar unpiloted probes as precursors to the establishment of a lunar base. These missions, possible with existing technology, would provide the detailed surface mapping necessary for selecting a site for the proposed base. Building the base as efficiently and economically as possible will require a new generation of space hardware, however. The agency's Moon strategy assumes that all vehicles for the lunar mission would be constructed, fueled, and maintained in low Earth orbit, in the neighborhood of the space station, to avoid the expense of having to escape Earth's gravity each time. NASA thus envisions the development of a new heavy lift launch vehicle (HLLV) that would be capable of putting 200,000-pound payloads—well beyond the capacity of the shuttle—into LEO. Other workhorses would include orbital maneuvering vehicles (OMVs) for transfer activities around the space station and orbital transfer vehicles (OTVs) to move payloads between low Earth orbit and the Moon. All of the proposed craft would employ chemical rockets.

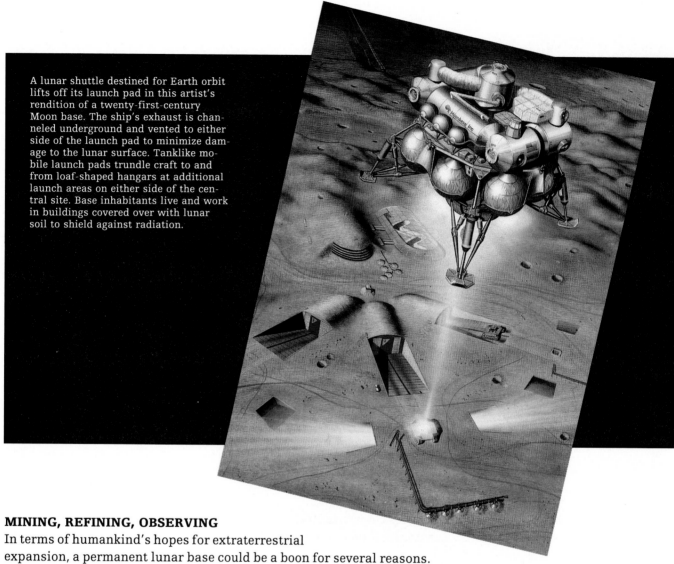

A lunar shuttle destined for Earth orbit lifts off its launch pad in this artist's rendition of a twenty-first-century Moon base. The ship's exhaust is channeled underground and vented to either side of the launch pad to minimize damage to the lunar surface. Tanklike mobile launch pads trundle craft to and from loaf-shaped hangars at additional launch areas on either side of the central site. Base inhabitants live and work in buildings covered over with lunar soil to shield against radiation.

MINING, REFINING, OBSERVING

In terms of humankind's hopes for extraterrestrial expansion, a permanent lunar base could be a boon for several reasons. For example, the Moon is an untapped source of oxygen, the breath of life to spacefarers and, in liquid form, one of the primary fuels of chemical rockets. Lifting large quantities of oxygen up from Earth to supply the space station and a fleet of spacecraft would be considerably more costly than mining it on the lunar surface and lifting it out of the Moon's lesser gravitational field— just one-sixth as strong as Earth's. As it happens, oxygen, bound up with other substances, constitutes about 40 percent of the Moon's soil by weight and can be extracted by straightforward mining and refining techniques. In addition, lunar mining may yield other valuable substances such as titanium, aluminum, iron, and magnesium.

The Moon also promises to be an exceptional place to study the universe. Lunar orbit is free of the spaceborne detritus that endangers astronomical instruments in orbit around Earth, and, with an entire atmospheric mass that amounts to less than the air inside New York's Madison Square Garden, it has no blanket of atmosphere to obscure or distort the view. Moreover, the Moon's low gravity would allow easy construction of large observational instruments, such as a proposed 16-meter optical telescope and a fully steerable 500-meter radio telescope.

Best of all, the lunar surface is extremely stable: A typical seismic event on the Moon is 100 million times less energetic than its average counterpart on Earth, moving the ground only about one-billionth of a meter. This seismic stability would permit the establishment of immense, electronically linked arrays of many telescopes that would have the resolving power of a single telescope miles in diameter. A proposal for such an array, known as LOUISA (Lunar Optical-Ultraviolet-Infrared Synthesis Array), envisions two concentric rings of 1.5-meter telescopes that would, in principle, be able to resolve a dime on the surface of Earth. LOUISA could detect Earth-like planets orbiting stars in the galactic neighborhood. A similar array of radio telescopes could perform comparable wonders. The Moon would also be an ideal place for observations of cosmological interest, such as the detection of hypothetical gravity waves and cosmic "strings" that may have formed in the aftermath of the Big Bang.

The establishment of such observatories would not necessarily require a long-term human presence on the Moon. One NASA astronomer has suggested, for example, that initially, small robotic observatories could be gently deposited on the lunar surface without on-the-scene human intervention. He is not alone in assuming that the field of robotics, already well established, will have come sufficiently far by then to make this feasible. Many scientists assume that robotic devices will be capable of performing routine maintenance even on later, more complex installations.

Humans will have a major role to play on the Moon, however, simply by proving that they can survive there. Lunar mining operations to produce raw materials for construction on the Moon and in Earth orbit could be, like the observatories, largely automated. But the construction of the lunar base itself is likely to require direct human participation. And while robots may be helpful in prospecting, geologist-astronauts will be essential for on-site evaluations of lunar resources. In particular, geologists would like to investigate whether the Moon possesses any deposits of ice that could support the water needs of the Moon base.

A RUGGED OUTPOST

Life on the Moon will probably resemble life in a nuclear submarine most of the time. The base itself is likely to consist of connected cylindrical modules buried under about six feet of lunar topsoil to protect people from solar radiation and unpredictable solar flares, violent outpourings of high-speed particles and x-radiation. With no clouds to block any of the Sun's rays, it will be dangerous for lunar colonists to spend much time aboveground, but solar energy collectors arrayed on the surface will supply the base with much of its power. Alternate energy sources, such as small nuclear reactors, would also be necessary, of course, inasmuch as the lunar night lasts two weeks.

As the base grows, operations will become more routine. The oxygen-extraction plant, for example, will send regular deliveries to facilities in Earth orbit, providing them with all the requirements for life support and propulsion. Crews assigned to the Moon will spend six months to a year at the base before rotating home. Meanwhile, the base will also support a transient population of visiting astronomers, geologists, low-gravity researchers, and other scientists. Investigators will look into ways of using helium-3, extracted from the lunar soil, to fuel fusion reactors. The isotope, a light form of helium that is borne on the solar wind, is rare on Earth, which is shielded by its magnetosphere, but relatively plentiful on the exposed lunar surface. If fusion reactors become practical in the next century, lunar helium could become an important export item from the Moon base.

All of this activity will require the development of a vast extraterrestrial infrastructure. The proposed heavy lift launch vehicles would make it more economical to boost people and payloads from Earth to the facilities in low Earth orbit, from whence fleets of transfer vehicles would routinely shuttle to and from the Moon. Eventually, lunar resources may be used to construct larger and more versatile space stations, or even an immense array of energy-generating satellites.

A SOLAR COLLECTOR

More than twenty years ago, in 1968, Czechoslovakia-born engineer Peter Glaser of Arthur D. Little, an international management and technology consulting firm headquartered in Cambridge, Massachusetts, came up with the idea to supply Earth's power needs by the orbital capture of solar energy. Specifically, he proposed that a large satellite in geostationary orbit 22,300 miles over the equator could collect solar energy, convert it to microwaves, and then beam it down to receiving antennas on the ground. The idea seemed technically feasible, and it was appealing enough that the Department of Energy spent nearly $20 million investigating the concept in the late 1970s but ultimately balked at the estimated price tag of $781 billion.

When the National Academy of Sciences in Washington, D.C., reexamined Glaser's proposal a few years later, the projected costs almost quadrupled: The academy concluded that a solar power satellite array sufficient to provide for the nation's energy needs would cost on the order of $3 trillion over fifty years. Some other difficulties were identified as well, including the unknown effects of bombarding Earth with such a torrent of microwave energy, and the possible damage to the atmosphere caused by the staggering number of rocket launches that would be required to build the solar collectors. Glaser's grand vision was quietly put on the shelf.

A variation on the original concept still holds some promise, however. In July 1989, physicist David Criswell, a lunar resources specialist at the University of California in San Diego, suggested that Glaser's orbiting space platforms are unnecessary. According to Criswell, energy could be beamed to Earth from solar arrays stationed on the surface of the Moon. Although the effects of the microwave transmissions remain a potential problem, a NASA study group has concluded that, in the long run, the Moon could play a major role in supplying terrestrial electricity.

In the very long run, the Moon base could grow into a small city, with a permanent population and a local economy spurred by the arrival of tourists from Earth. Some visitors might come for their health: A lunar medical center might specialize in treating those with heart disease, high blood pressure, and other cardiovascular ailments who would benefit from time spent in the Moon's low gravity. For similar reasons, the Moon might even become a retirement community for the elderly, who would welcome the respite from their long, losing battle against Earth's gravitational pull.

TRAINING FOR MARS

One of the more intriguing prospects for the relatively near term is that the Moon is likely to play a vital role in the continuing exploration of other planets in the Solar System. NASA's planning for the future envisions the Moon as a necessary way station on the road to Mars, for example. Much of the same infrastructure developed to support the Moon base would be essential for any Mars mission, and the base itself could be a training ground for Martian pioneers and perhaps the source of the liquid oxygen needed to fuel the expedition. In fact, the first human crews to go to Mars might depart for the Red Planet directly from the Moon or from lunar orbit.

Building on that experience, the lunar facilities might someday become the hub of all future human forays into the inner Solar System. In considering the quartet of planets nearest the Sun, scientists sometimes speak of the "Goldilocks Problem." From a human perspective, only Earth—having managed to maintain a mean global temperature above the freezing point of water and well below its boiling point—is neither too cold nor too hot but just right.

Prospects for human settlement of Venus and Mercury, both inboard of Earth, are bleak. Mercury, which is scarcely 36 million miles from the Sun on average, is a barren, rocky body where temperatures range from 800 degrees Fahrenheit at high noon—hot enough to melt zinc—to −300 degrees Fahrenheit during the eighty-eight-day-long Mercurian night. The United States space program's *Mariner 10,* the only craft that has visited Mercury, returned hundreds of images during the early 1970s that revealed a surface pocked with immense meteor-impact basins ringed by towering cliffs. Some scientists have speculated that in the perpetual shadows of those cliffs, primordial ice might still exist. If the ice is there, it could be a source of water for life support and rocket fuel for any human travelers who set foot on Mercury's unfriendly ground. No nation currently has plans to send another probe to Mercury, much less a manned mission, but an automated rover may someday be dispatched to analyze the composition of the planet's surface. And the small world would certainly be a useful site for an automated solar observatory. Such a facility could provide lifesaving warnings of the onset of dangerous solar flares.

Like Mercury, Venus is a world better suited to robots than humans. Located 67 million miles from the Sun on average, it is similar to Earth in size and mass but is blanketed in a dense atmosphere of carbon dioxide, which is responsible for the heat-trapping phenomenon known as a runaway greenhouse effect. The surface temperature is more than 900 degrees Fahrenheit, and sulfuric acid droplets pervade the opaque cloud layers. Despite the blistering conditions, between 1967 and 1984 a score of Soviet and American probes survived for short periods in the atmosphere and on the surface. In 1990, the American *Magellan* spacecraft went into orbit around Venus and began producing detailed radar-imaging maps of its terrain. Among *Magellan*'s initial findings were large faults in the planet's crust and numerous small volcanoes; scientists have yet to determine whether these indicate that the planet is still geologically active. In the future, the development of advanced robotic technology may enable humans to establish a "telepresence" on Venus for research purposes, but no human landing is likely to be attempted in the foreseeable future.

NO VACATION SPOT

Mars, the fabled Red Planet, is a much more Earth-like world than either Mercury or Venus, but it nevertheless has a forbiddingly hostile environment. On average, it orbits one and a half times farther from the Sun than does Earth and receives only 43 percent as much solar energy. The Martian atmosphere is mainly carbon dioxide with traces of nitrogen, argon, and oxygen. In addition, it has a pressure of just .007 that of the terrestrial atmosphere—the equivalent of the pressure atop an earthly mountain 100,000 feet high, more than three times the height of Mount Everest. The combination of great distance from the Sun and a very thin atmosphere makes Mars a frigid world, with an average global temperature of about minus sixty degrees Fahrenheit. Lacking an atmosphere with a protective layer of ozone, the planet is bathed in harsh ultraviolet radiation, which can cause, among other things, skin cancer and retinal damage in unprotected humans, as well as genetic damage in plants. Despite all these drawbacks, American space scientist Bruce Murray has likened Mars to "a Club Med site compared with any other locale in the Solar System."

Accordingly, a number of earthly vessels will be sent to Mars in the next few years. They will not be the first. In the summer of 1976, two United States Viking landers touched down on the surface of the planet and carried out experiments designed to detect microbial life. The results were tantalizingly ambiguous, but most scientists concluded that there was no evidence that life had ever existed on Mars. However, the Vikings examined only two rather bland locations, and some researchers insist that the question of Martian life is still open. As one science writer has put it, "Viking asked Mars if there was life there, and Mars replied, 'Please rephrase the question.'"

Mars Observer, the first American mission to the Red Planet since Viking, is scheduled to arrive in Martian orbit in early 1993. Basically a mapping

In August 1990, the *Magellan* spacecraft, seen here as it was deployed from the cargo bay of the shuttle *Atlantis* on May 4, 1989, began sending back detailed radar images of the surface of Venus, Earth's nearest planetary neighbor and a possible site for robotic exploration. The test image at left covers a strip twelve miles wide and ninety miles long of the volcanic upland region known as Beta Regio. Across the middle of the image, a fine network of valleys and ridges may be the product of tectonic activity. The smooth dark patch at the top was probably formed by lava flows recent enough that the region has not been scarred by meteoritic impact cratering.

Though found on a planet more than 52,000,000 miles away, this early morning scene—a photograph taken near Chryse Planitia on Mars by the *Viking 1* lander—evokes images of desert areas in California and Mexico, giving scientists from around the world a basis for preparing upcoming exploratory missions to the Red Planet.

mission, the probe will chart the mineralogical, physical, and chemical properties of Mars and also search for signs of water. At one time a great deal of water flowed there, perhaps planet-bathing oceans of it, but liquid water cannot now exist on the surface because of the present-day low atmospheric pressure. However, intriguing and controversial evidence of Martian "oases" has been found in the form of radar indications of near-surface liquid water, and some scientists believe that much of the ancient Martian endowment of water could now survive in subsurface layers. Water ice is known to exist in the polar caps, and may also be present in widespread subsurface permafrost. Water, in any form, would be essential to the existence of Martian life—past, present, or future.

ON-SITE SAMPLERS

If *Mars Observer* does not find water on Mars, future missions may. Both the Soviets and the Americans have a series of robotic Mars missions planned for the 1990s and beyond. Early in the next decade, there may be a *Mars Rover/ Sample Return Mission*, in which an automated vehicle will roam the rust-red deserts, collecting samples, making observations, and performing experiments. The *Rover* itself may be a conventional four-wheeled "buggy," or it may take a more exotic form. One interesting proposal, developed by graduate students at the University of Arizona's Lunar and Planetary Laboratory, is known as the Mars Ball, two large wheels connected by an axlelike hub carrying the control system and instrumentation *(page 38)*. The vehicle would roll forward as individual gasbags making up each wheel were deflated and then reinflated in a continuous cycle. The advantage of the Mars Ball is that it could safely negotiate rugged terrain that would halt a more conventional, less pliant land machine.

Another way to deal with the rocky Martian landscape is to fly above it. A number of scientists have toyed with the concept of exploring the Martian surface by means of the most ancient of all airborne vessels, the balloon, and a Soviet/French/United States team is actually designing one. The large helium-filled balloon would drift on the Martian winds, covering more than a hundred miles per day in the course of its ten-day mission and recording various kinds of information about the planet. Because the vessel would be at the mercy of the winds, directed navigation would be impossible, but the probe could carry out scientific observations at scores of sites. For more controlled exploration, a solar-powered, remote-controlled "Marsplane" might be possible. Or perhaps a hybrid, blimplike airship design might work best, allowing control while still offering the advantages of buoyancy. Such ingenious devices could greatly facilitate the exploration of Mars, whether by machines or by humans.

Eventually—perhaps by the year 2005—human beings will make their first landing on the Red Planet, paving the way for the establishment of a permanently occupied base. NASA has sketched a number of scenarios for such missions to Mars. In one scheme, known as a split-sprint mission, a vessel

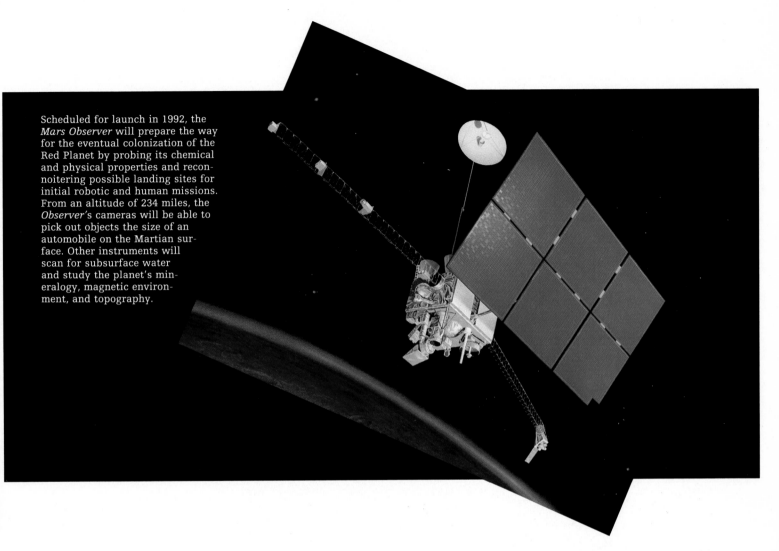

Scheduled for launch in 1992, the *Mars Observer* will prepare the way for the eventual colonization of the Red Planet by probing its chemical and physical properties and reconnoitering possible landing sites for initial robotic and human missions. From an altitude of 234 miles, the *Observer*'s cameras will be able to pick out objects the size of an automobile on the Martian surface. Other instruments will scan for subsurface water and study the planet's mineralogy, magnetic environment, and topography.

carrying only cargo would leave first, departing from Earth orbit. It would contain fuel for the return trip, a variety of gear for planetary exploration, and a two-part vehicle for descent and ascent. Because high speed would not be essential, the cargo vessel could economize on fuel by taking a slow, low-energy trajectory that would put it at its destination in a year or so. Once this ship signaled its safe arrival in Mars orbit, a second vessel would carry astronauts out to Mars by a more direct route that could be traversed in eight months. Crew and cargo would rendezvous in orbit around Mars, and some of the astronauts would descend to the surface for a stay of several weeks. They would then return to orbit and speed back to Earth in a six-month sprint. Another possibility is to leave from lunar orbit in a single vessel (fueled with Moon oxygen), spend eight months traveling to Mars, and stay on the surface of the Red Planet for about a year.

Although NASA has consistently favored a step-by-step, conservative approach to sending people on a Mars mission, other groups have advanced more radical ideas. One scenario, named "Mars Direct," was proposed in the spring of 1990 by the team of Robert Zubrin and David Baker, engineers at Martin Marietta Aerospace. Unlike NASA's plan to develop Earth orbit or lunar orbit as staging areas for future Mars missions, Zubrin and Baker's proposal calls for a heavy lift launch vehicle, based on space shuttle technology, that would use the upper stage of the rocket to send payloads of up to fifty-two tons on a direct path to Mars.

The first launched payload, which would arrive at Mars in 1997, would be an unpiloted, unfueled, two-stage vehicle for returning to Earth from the Martian surface. Parachuted to the ground, this craft would be crammed with 6.4 tons of liquid hydrogen, a life-support system, food, a 100-kilowatt nuclear reactor, an automated chemical plant, a utility truck, and a set of robotic rovers. Powered by the reactor, the chemical factory would combine the liquid hydrogen with carbon dioxide from the Martian atmosphere to produce methane and water. Hydrogen and oxygen could then be derived from the water; the oxygen would go into storage, while the hydrogen would be recycled. In this way, the factory could produce about 120 tons of methane-oxygen rocket fuel in approximately a year—enough to fill up the tanks of the Earth return craft and to support on-the-ground exploration. As Zubrin noted, "This may sound somewhat involved, but actually the chemical processes employed are nineteenth-century technology."

Two years after the first launch (and assuming that mission controllers back on Earth had determined that fuel production was completed as planned), the next flight would depart from Earth. This one would carry a crew, but would also be accompanied by an unpiloted craft identical to the first. After a journey of six months, the crew would land near the waiting and fully fueled return vehicle and spend as long as 600 days exploring the surface. The new return vehicle–automated fuel factory, meanwhile, would land several hundred miles away. Two years later, a second crew would land at this site while a third factory landed several hundred miles away. Thus,

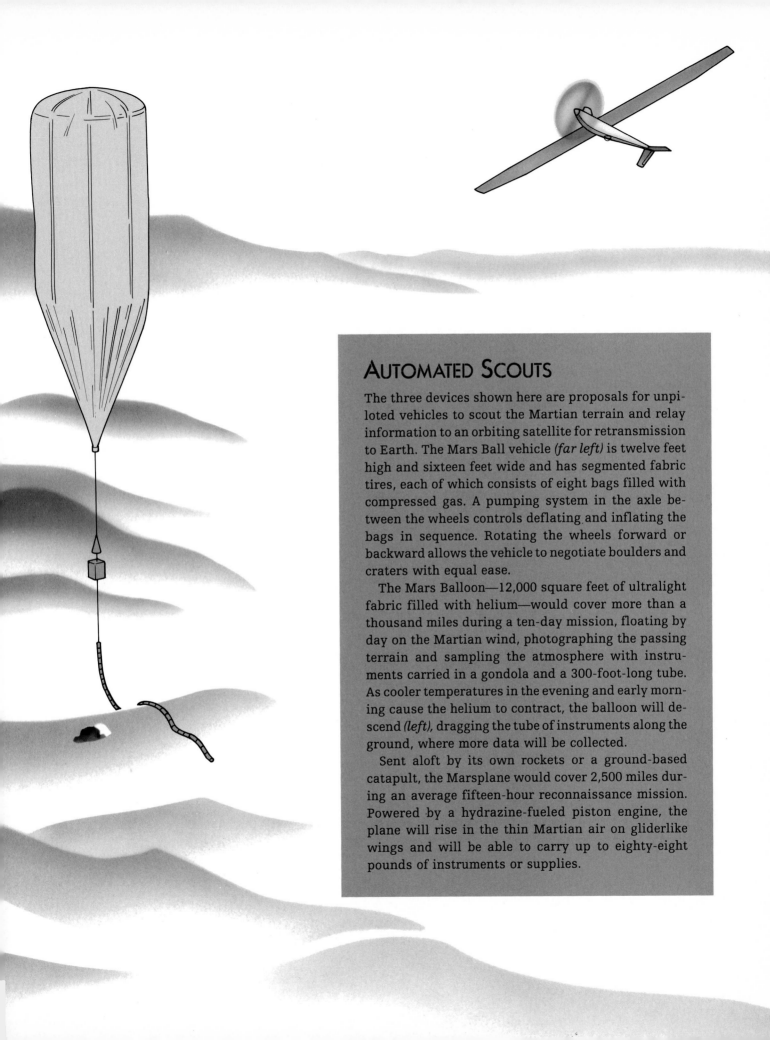

AUTOMATED SCOUTS

The three devices shown here are proposals for unpiloted vehicles to scout the Martian terrain and relay information to an orbiting satellite for retransmission to Earth. The Mars Ball vehicle *(far left)* is twelve feet high and sixteen feet wide and has segmented fabric tires, each of which consists of eight bags filled with compressed gas. A pumping system in the axle between the wheels controls deflating and inflating the bags in sequence. Rotating the wheels forward or backward allows the vehicle to negotiate boulders and craters with equal ease.

The Mars Balloon—12,000 square feet of ultralight fabric filled with helium—would cover more than a thousand miles during a ten-day mission, floating by day on the Martian wind, photographing the passing terrain and sampling the atmosphere with instruments carried in a gondola and a 300-foot-long tube. As cooler temperatures in the evening and early morning cause the helium to contract, the balloon will descend *(left)*, dragging the tube of instruments along the ground, where more data will be collected.

Sent aloft by its own rockets or a ground-based catapult, the Marsplane would cover 2,500 miles during an average fifteen-hour reconnaissance mission. Powered by a hydrazine-fueled piston engine, the plane will rise in the thin Martian air on gliderlike wings and will be able to carry up to eighty-eight pounds of instruments or supplies.

in a series of leapfrogging deployments—and using the chemically fueled rovers in each mission—human explorers would become acquainted with much of the surface of Mars.

Initially, they would concentrate on such investigations as looking for water and for life, living or fossilized, and making detailed observations of the Martian climate and geology. As the habitations grow, however, follow-on pioneers will engage in a wide variety of activities familiar to residents of Earth: building, farming, mining, manufacturing, and so on.

Meanwhile, travel between Earth and Mars will become routine. American engineer John Niehoff has proposed that several spaceships could be placed in elliptical orbits that would cause them to repeatedly pass near the two planets. At low expense, these craft could carry a steady stream of passengers and cargo across the ocean of space separating the worlds, swept along their curving courses by gravity alone.

This idea is being championed by former astronaut Buzz Aldrin, and was also cited in "Pioneering the Space Frontier," the 1986 report of the National Commission on Space. Aldrin envisions a pair of reusable craft that he calls Cyclers, which would take advantage of the gravitational slingshot effect gained from precisely directed flybys of planets, in this case Earth and Mars. One craft would be used for outbound journeys, the other for returns. Smaller vessels would rendezvous with the Cyclers to shuttle passengers down to and up from the surface of each planet.

By that time, presumably, many other vessels would be plying the Solar System, making round trips to the asteroid belt, the moons of the outer planets, and the borderlands of our stellar domain. Earth's Moon and even Mars will be familiar territory, transformed into human homes as the migrants move on. As Robert Zubrin expressed it, echoing Konstantin Tsiolkovsky's words of seventy years ago, "We've gotten too big for the cradle."

Thus far, every foray into the ocean of space has been accomplished by rocketry—the burning of chemical fuels to boost payloads up from Earth, to delicately adjust spacecraft orbits and trajectories, and to slow vehicles down as they approach some distant landfall or return home again. While rockets will continue to power spaceflight for decades to come, many astronautical engineers believe that the path to the stars will widen to include transport based on ancient principles embodied in the bola or sling.

The implements at the heart of this vision are space tethers—lightweight lines, some only a few millimeters in diameter, made of ultrastrong synthetic fiber. Joining objects across a distance of tens or hundreds of miles, these tethers would be able to perform feats of momentum transfer that seem almost magical. For example, a satellite unreeled on a tether from a space shuttle will jump to a higher orbit when released, at the same time causing the shuttle to drop back toward Earth, all with a minimal use of rocket fuel. More futuristically, a spacecraft en route to Mars could make a midcourse correction with no expenditure of fuel by spearing a passing asteroid with a specially tipped tether and, in crack-the-whip fashion, swinging off in a new direction.

The simple but revolutionary notion of using tethers in space was developed in the 1970s and early 1980s by the Italian engineer Giuseppi Colombo, a professor at the University of Padua. Today, scores of researchers at NASA, the Jet Propulsion Laboratory in Pasadena, California, and other centers of space science in Italy are exploring the potential of these high-tech ropes. As shown on the following pages, their ideas not only embrace a range of fuel-saving maneuvers but even include plans for electrically conductive tethers that could turn Earth's magnetic field into a power supply.

TAUT LINKS TO THE SPACE STATION

A fixture of most plans for the outward expansion of humankind is a space station in permanent Earth orbit. Many designs for such a station have been proposed, and some of the most elegant incorporate the fast-developing technology of space tethers. Held taut by competing centrifugal and gravitational forces *(box, right)*, a tether might tie together an array of living quarters, laboratories, and docking facilities. All would orbit as a unit, yet would have the advantage of environmental independence. The arrival of a space shuttle, for example, might jolt the docking platform, but the vibrations would not disrupt delicate telescopes in a tethered structure floating miles away.

The space station pictured here would involve a sixty-mile-long tether linking three structures: a combination dock-and-hangar with four propellant tanks, an elevator, and the main space station, seen at right in the distance. Spacecraft bound for Mars could be assembled in the hangar with the aid of a robot arm, and then take on fuel, a process made easier by the centrifugal force that the tether hookup would engender. People, supplies, and spacecraft components would shuttle between the upper platform and the main space station in the electrically powered elevator, shown here about midway along the tether. The elevator might also function as a laboratory for microgravity experiments: Gravity could be reduced to zero or increased by small degrees simply by moving the elevator to different positions along the tether.

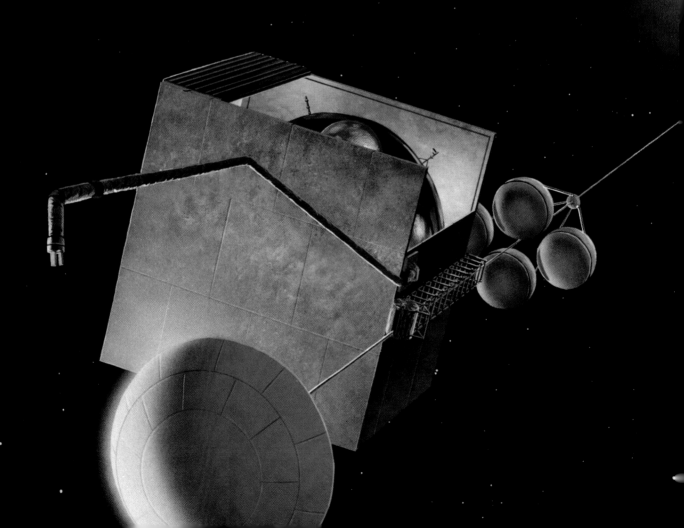

Free orbit. Celestial mechanics dictate that, when two free-flying objects orbit at different altitudes, the lower one moves faster than the upper one and will complete its orbit in a shorter time.

Tethered orbit. Two linked objects are, in effect, forced to compromise on a single orbital velocity by speeding up the higher object and slowing down the lower one. The centrifugal force on the higher satellite counteracts the gravitational force tugging on the lower one to keep the tether taut.

43

Transferring momentum. When the tether linking two orbiting satellites is disconnected, the inner object, which is moving slower than normal, falls into a lower, elliptical orbit whose high point, or apogee, is the point of release. For the outer object, which is moving faster than normal, the low point, or perigee, of its new, higher orbit is the tether release point. The change in altitude for each is seven times the original distance from the object to the center of mass on the tether.

Orbital Transfers

In the not-too-distant future, space shuttles may use tethers to boost payloads into higher orbit and simultaneously send themselves homeward by a controlled parting of the ways, as illustrated here with a space probe bound for Mars. The probe is being slowly unreeled from the shuttle on a reusable tether that will be nearly 200 miles long at the point of release. Small thrusters on the probe separate it from the shuttle for the first mile or so; thereafter, centrifugal force will carry it away.

The linked system's center of mass—located at a point along the tether where the competing centrifugal and gravitational forces exactly offset each other—will travel at the right speed for its orbital altitude. But the shuttle will be traveling slower and the space probe faster than if they were flying free. When released from the tether, the probe, experiencing a greater centrifugal than gravitational force, will begin to rise. The shuttle, experiencing a greater gravitational than centrifugal force, will begin to drop. A rule of thumb for space-tether technology says that objects will rise or fall seven times the distance between the object and the center of mass on the tether. The result would be an impressive savings of fuel—on the order of 7,000 pounds for the shuttle and an average-size probe—for both the lifting of the payload and the shuttle's descent to Earth.

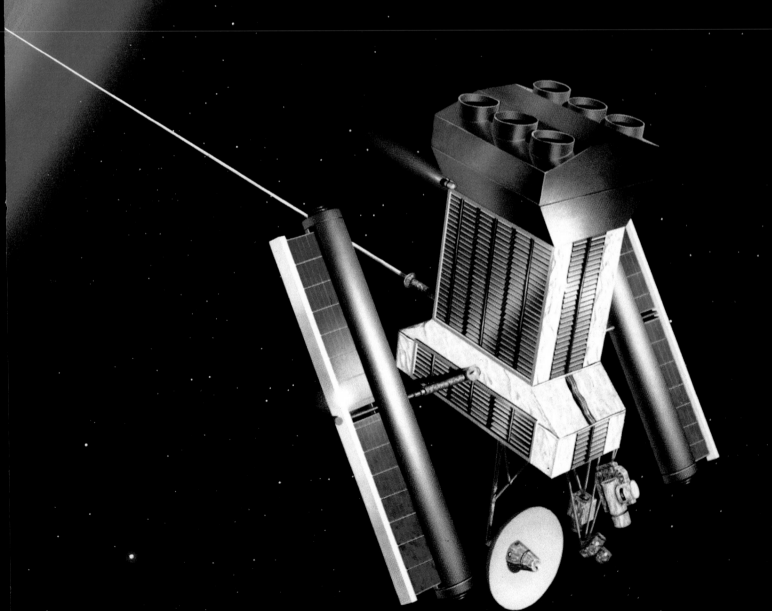

A Tethered Dynamo

Along with their ability to mediate gravitational and centrifugal forces, orbiting tethers may someday generate electricity for spacecraft, or even provide thrust, by converting electrical energy into orbital energy and vice versa through interactions with Earth's magnetic field. The electrodynamic principle is familiar: Moving a wire circuit through a magnetic field causes current to flow; reciprocally, running current through the wire creates a magnetic field.

Unlike an ordinary space tether, the electrodynamic version would be insulated and made of conductive material. To complete the circuit necessary to generate a current, the tether would have to be linked electrically to Earth's ionosphere, the rarefied upper region of the atmosphere where solar radiation has stripped atoms of their electrons. In the speculative scene at left, a conductive tether stretches six miles above and below a space station orbiting Earth at an altitude of 350 miles. At each end, special terminals heat an inert gas to create a plume of plasma—a soup of charged atomic particles that allows ionospheric electrons to flow into and out of the tether *(box),* either to supply power to the station or to generate thrust for orbital adjustments.

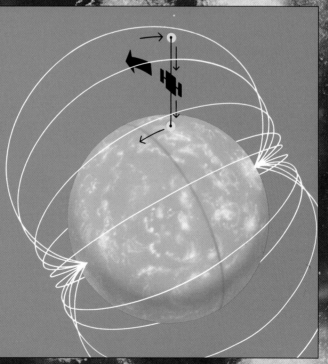

Orbiting dynamo. As an electrodynamic tether attached to the space station moves through Earth's magnetic field, electrons in the ionosphere are attracted to an invisible stream of plasma issuing from the upper terminal; they flow into the tether, down its length, and return to the ionosphere by way of plasma issuing from the lower terminal, thus completing an electrical circuit. The resulting current generates a magnetic field whose polarity is opposite that of Earth's magnetic field, producing resistance that slows the space station and causes it to lose altitude. If the space station expends power to reverse the direction of the current through the tether, the field's polarity will match Earth's, boosting the space station's speed and causing it to rise.

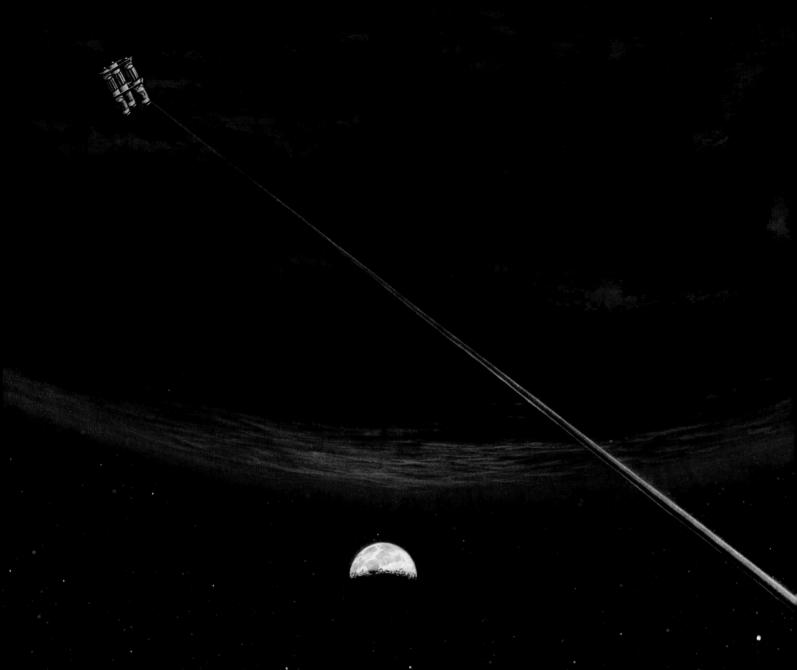

A Rotating Skyhook

Of the many uses envisioned for space tethers, one of the most dramatic is a gigantic version of the sling David used to slay Goliath. Here, such a sling is depicted hurling a spacecraft *(far right)* from low Earth orbit into space to begin a longer journey. Known to tether researchers as a rotating skyhook, this theoretical system consists of a sixty-mile-long tether running from a massive platform (in the distance at upper left) to a capture-and-release mechanism that can handle a variety of payloads *(right)*. The mechanism, which rotates around the platform like a giant pinwheel, snares space vehicles or cargoes sent up from Earth and whips them to escape velocity for fuel-free boosts to the Moon and points beyond.

Because of the transfer of momentum involved, the skyhook loses orbital altitude each time it throws a payload into space. Fortunately, the loss can be made good by capturing incoming payloads returning to Earth from, say, Mars. Each capture of an incoming mass would transfer momentum to the platform and boost it back to a higher orbit to continue its work.

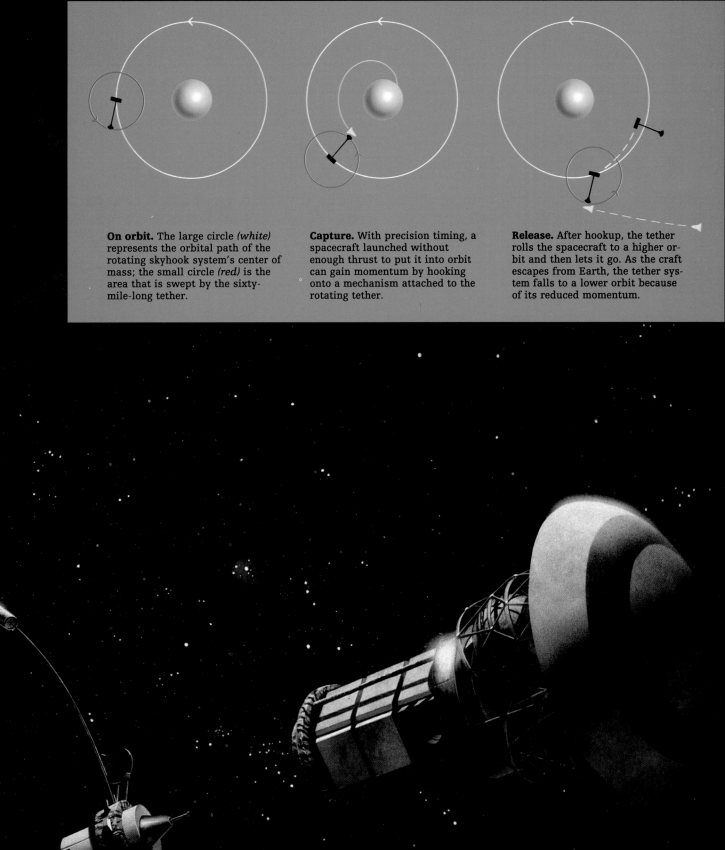

On orbit. The large circle *(white)* represents the orbital path of the rotating skyhook system's center of mass; the small circle *(red)* is the area that is swept by the sixty-mile-long tether.

Capture. With precision timing, a spacecraft launched without enough thrust to put it into orbit can gain momentum by hooking onto a mechanism attached to the rotating tether.

Release. After hookup, the tether rolls the spacecraft to a higher orbit and then lets it go. As the craft escapes from Earth, the tether system falls to a lower orbit because of its reduced momentum.

AN ARTIFICIAL GRAVITY-ASSIST

Numerous space probes dispatched into the Solar System on reconnaissance missions have taken advantage of the powerful gravitational tug of planets or moons to alter course and gain speed. Asteroids, far more numerous, cannot be used for such slingshot turns because their gravity is so weak. But specially tipped tethers may someday make possible artificial gravity-assists by briefly connecting spacecraft to the rocky bodies that orbit the Sun by the thousands.

In the scene at left, a vehicle bound for Mars fires a harpoonlike tether across 200 miles of space toward a receiving dish left on an asteroid by an earlier mission. Once the tether is anchored, the craft becomes a ball on a string, whirling in an arc around the asteroid and picking up speed. At exactly the right moment, a signal from the spacecraft frees the tether, which is then reeled in for reuse. The maneuver could cut fuel consumption on a trip from Earth to Mars by 30 percent.

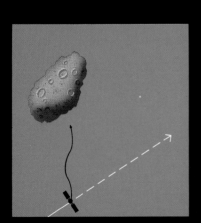

Gravity-assist. A Mars-bound craft, traveling on a path marked by the white line above, fires a tether toward an asteroid.

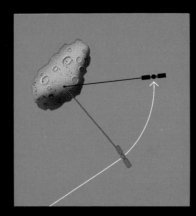

Detour. When the tether attaches to the asteroid, which has a mass millions of times that of the craft, the craft is pulled off its course.

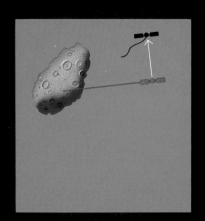

New heading. After swinging around the asteroid, the vehicle releases the tether and, at increased speed, heads off on a new course.

2/Pioneers and Settlers

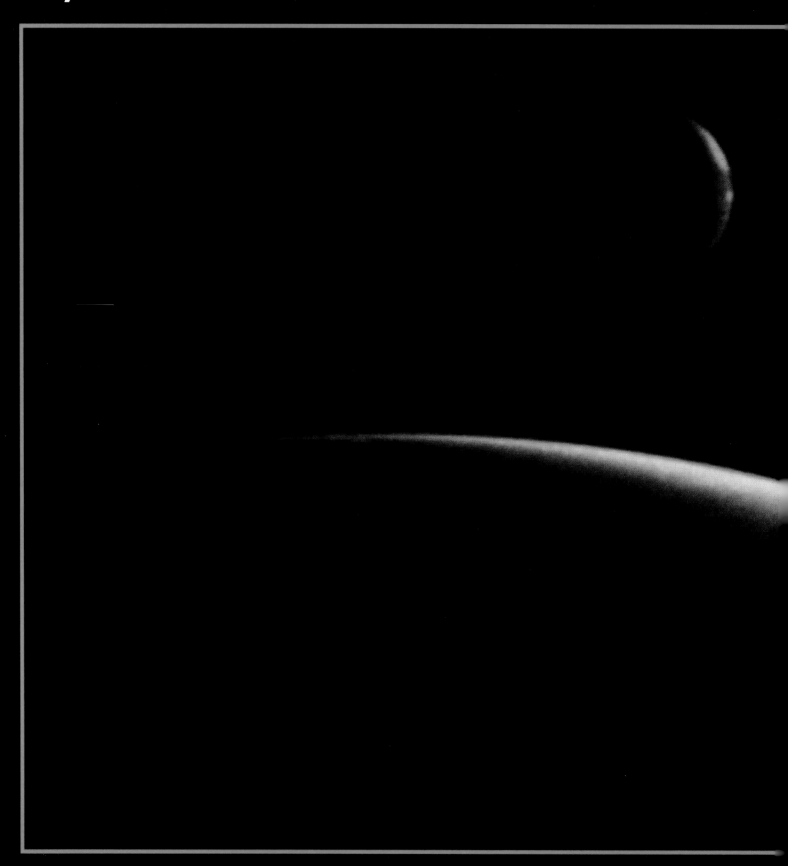

Almost three billion miles from the
Sun, the moon Triton rises over
Neptune in one of the last images
transmitted by *Voyager 2* in 1989
as the probe headed for the edge of
the Solar System and the interstellar void beyond. The gaseous planet and its satellites are the most
distant bodies yet visited by a
spacecraft from Earth.

or some time now, the inhabitants of Asteroid Resource Colony I have been filtering into the central arena, arriving by electromagnetic shuttle from four habitation pods that rotate slowly about the colony's hub to provide artificial gravity in the living quarters. As the settlers reach the hub, they enter a virtually weightless environment and experience a feeling of buoyancy that heightens their already festive mood. The arena, covered by a spacious dome, normally hosts colony council meetings, as well as matches of low-gravity "sockey"—a hybrid of soccer and hockey contested with sticks and balls on a three-dimensional field studded with free-floating goals. Today, draped with bunting and abob with balloons, the enclosure resembles nothing so much as a convention center on Earth. In fact, a limb of that cloud-swirled planet can be glimpsed dimly through opaque panels treated to prevent dangerous solar radiation from penetrating the dome. The sight moves some in the crowd to nostalgia for their distant home. Others are filled with a profound sense of pride, stemming from the knowledge that they need no longer consider themselves bound to one planet alone.

ARC I, as the residents refer to their colony, represents the fullest flowering of twenty-first-century space technology. The rotating assemblage was fabricated over the course of several years on a mile-wide asteroid that a colony preparation team captured as its looping orbit around the Sun brought it near the planet Earth. Bivouacked in a hollowed-out cave, the team deployed a mass driver, a propulsion engine fueled by pulverized rock drilled from the asteroid itself. By flinging the debris outward from the surface at high speeds, the mass driver slowly nudged the asteroid out of its solar orbit and into Earth orbit. Bolstered by support crews, the advance team then began the task of refining asteroidal ore in solar furnaces to yield iron, nickel, copper, and other elements. From these, the crews forged structural components for the central arena—built at an elevated point on the asteroid—and the four cylindrical habitats tethered to the hub by arrays of cable. The most sophisticated habitat of its kind in Earth orbit, with abundant mineral resources remaining to be tapped, the colony soon lured hundreds of pioneers from a planet all but devoid of frontiers to tame and wealth to exploit.

Indeed, the event that brings the jubilant colonists together in the hub now is one that has ceased to be an occasion for general rejoicing on a desperately crowded Earth. With quiet dignity, a young woman assumes the place of honor beneath the crown of the dome and, ringed by appreciative settlers, effortlessly raises her infant skyward—a baby girl, named Astra, the first child to be born on the asteroid colony and, in the minds of many who are commemorating her birth, the first among them who has the right to claim kinship with the stars. . . .

PROPHETIC FLIGHTS

. . . Like the earliest promises that humans would someday walk on the Moon, predictions that pioneers will one day abandon Earth for self-sustaining orbiting habitats have a fantastic ring to them. Dreams of colonizing space have fueled many a fiction, and those who attempt to think realistically about the possibility and predict what form it might take risk being regarded as mere storytellers. Yet experience shows that imaginative flights by the visionaries of one generation can chart the way for serious investigation by their successors.

In 1929, for instance, the British crystallographer J. D. Bernal, one of the founding fathers of molecular biology, published a tract entitled *The World, the Flesh and the Devil,* in which he suggested that spacefarers might one day land on asteroids and exploit their resources in subterranean shelters—musings that doubtless raised more than a few eyebrows among Bernal's scientific peers. Within a few decades, however, the prospect of mining asteroids was a legitimate topic of discussion at conclaves devoted to space exploration. Interest in asteroids was spurred by the realization that, together with the Moon, they offer virtually all the raw materials needed to build large orbiting colonies, a construction job that would be prohibitively costly if engineers had to overcome Earth's gravity to boost components into space. And the experience garnered by such colonies will be vital to the future of spaceflight. For as humans venture ever farther from the home planet and set their sights on the stars, return journeys may cease to be practical or even desirable. The supreme voyages will likely be one-way excursions, carried out by men and women conditioned to life in enclosed, self-sustaining habitats and equipped to colonize distant worlds.

Scientists began to look seriously at asteroids as targets of exploration in the 1960s, when plans for a lunar landing fostered a climate in which one-time space fantasies suddenly emerged as realistic possibilities. Leading the way was Dandridge Cole, an analyst with General Electric's Missile and Space Division. Several years before astronauts first walked on the Moon, Cole was plotting the next step for the space program. He argued forcefully that piloted missions to some of the minor planets—the preferred scientific term for asteroids—should precede an expedition to the major planet Mars. The asteroids Cole had in mind were not those in the main belt between Mars and Jupiter, but members of a smaller group whose eccentric careers around the Sun lead them to approach, and in some cases cross, Earth's orbit. Cole noted that one such Earth approacher whose orbit was well known to astronomers, the minor planet Eros, would in 1975 pass within 14 million miles of Earth—

about one-third the distance to Mars. Reflecting the heady optimism of the times, Cole called for a mission to Eros to set the stage for a more ambitious venture to Mars in the 1980s.

ASTEROID ROUNDUP

For Cole, the lure of asteroids was not limited to their accessibility. In the book *Islands in Space,* on which Cole collaborated with space writer Donald Cox in 1964, the authors argued that minor planets were important "as a source of raw materials for supporting space exploration and colonization of our solar system." Their claim was based on the reasonable assumption that asteroids are similar in composition to meteorites, which tend to be either chunks of iron and nickel or stony fragments called chondrites, which contain an array of useful compounds, including water and hydrocarbons. (Subsequent analysis of the light and heat reflected by asteroids has confirmed a compositional link between meteorites and asteroids, many of which appear to be rich in carbon, others in metals.) Of course, efforts to mine asteroids would yield little if missions were confined to those rare occasions when minor planets of known orbit came close to Earth. The solution, the authors boldly suggested, was to land on promising asteroids and guide them into Earth orbit—a feat that might be accomplished, they added, by using megaton-range nuclear bombs for course corrections. Scientists who later pondered the problem of asteroid capture modified this plan, advocating some form of mass driver for more precise guidance *(pages 88-89).* But few doubted that the capacity to corral a minor planet would lie within the reach of twenty-first-century technology.

Among those who embraced the concept of exploiting asteroids to support space colonies was an astronomer and would-be astronaut named Brian O'Leary. In the early 1950s, while he was still in grade school, he had pored over magazine articles about space travel and dreamed of one day visiting the Moon. In 1967, it appeared that his fantasy might be fulfilled when he was selected by NASA to join the first group of scientists to undergo the rigorous astronaut-training program. Unfortunately for O'Leary, NASA insisted that the scientists prove their skills as jet pilots like the other trainees. He bridled at the requirement, convinced that flight school was "an unnecessary, distracting prerequisite for the trip into space." After his second solo flight, a harrowing experience in which he approached the runway a full forty miles per hour above the prescribed speed, he left the training program to return to the academic world.

But the idea of escaping Earth's grasp continued to fascinate O'Leary. Rejecting the notion that the Moon and other objectives in space were regions to be conquered and then abandoned like mountain peaks on Earth, he argued for a long-term program that would plant the seeds of human culture permanently on distant worlds. "We are today on the verge of a second Copernican revolution," he wrote in his book *The Fertile Stars.* "Freeing ourselves from the bondage of Earth's gravity may be no less profound a leap in the perception of ourselves and our planet than the original Copernican revolution was nearly 500 years ago.''

In O'Leary's view, efforts to colonize space had been hampered by the notion that Earth must always be the launch pad. Instead, he advocated a new concept: ''to launch from the Earth only what is needed in order to mine the Moon or the asteroids and take from them the materials required to build things in space, where there is constant solar energy.'' O'Leary estimated that an Earth-approaching asteroid the size of a football field would provide ''a million tons of material, enough to build a space habitat for 10,000 people.'' And he sketched out a mission to accomplish the task. A team of astronauts would depart from a base on the Moon or in low Earth orbit with mining equipment and a mass driver assembled in space, rendezvous with the asteroid several months later, and fasten cables to the spinning body to bring it under control. Crew members would then deploy the mass driver to propel the asteroid toward Earth orbit. En route, they would excavate a workshop where they could harness solar power to begin refining water, oxygen, and various minerals and metals.

Although it might take a year or two for the asteroid to reach its assigned orbit around the home planet, the time and effort would be well spent, O'Leary

An early prophet of expansion into space, engineer Dandridge Cole *(near right)* predicted in the 1960s that asteroid minerals would be a lucrative prize for future space ventures. Two decades later, astronomer Brian O'Leary suggested that these minor planets could supply Earth not only with a substantial portion of its raw materials but also, indirectly, with food—grown in orbiting structures built from asteroidal matter.

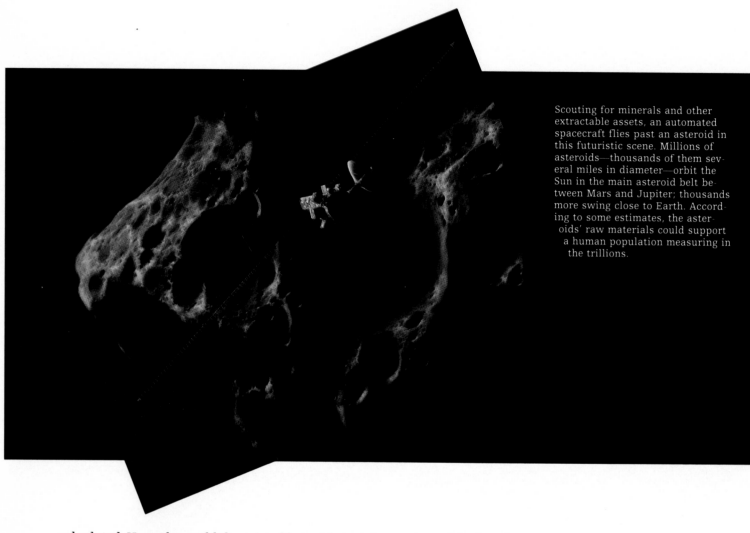

Scouting for minerals and other extractable assets, an automated spacecraft flies past an asteroid in this futuristic scene. Millions of asteroids—thousands of them several miles in diameter—orbit the Sun in the main asteroid belt between Mars and Jupiter; thousands more swing close to Earth. According to some estimates, the asteroids' raw materials could support a human population measuring in the trillions.

calculated. Not only would the asteroid provide certain raw materials that the Moon lacked, but those materials would be considerably easier to transport, given the asteroid's negligible gravity. Even the Moon, with its low escape velocity, exacts a measurable energy toll on vehicles lifting off its surface, and in a high-volume mining operation, this small surcharge could increase costs substantially. Some of the minerals extracted from asteroids would be in such demand on Earth that it would be profitable to ship them back to the planet. But the rest would go to build orbiting colonies whose inhabitants, once enclosed, would rely almost entirely on their own devices and on the renewable resources of their artificial worlds.

HABITATS ON A HIGH FRONTIER
O'Leary was encouraged in his vision of colonizing space by someone he had met during tryouts for the astronaut program, Gerard K. O'Neill. Like O'Leary, O'Neill never made it into orbit, but he continued to dwell imaginatively on the possibilities of life in space as a serious sideline to his work as a high-energy physicist at Princeton University. In 1969, O'Neill visited Cornell University, where O'Leary was teaching astronomy, and elaborated a lofty

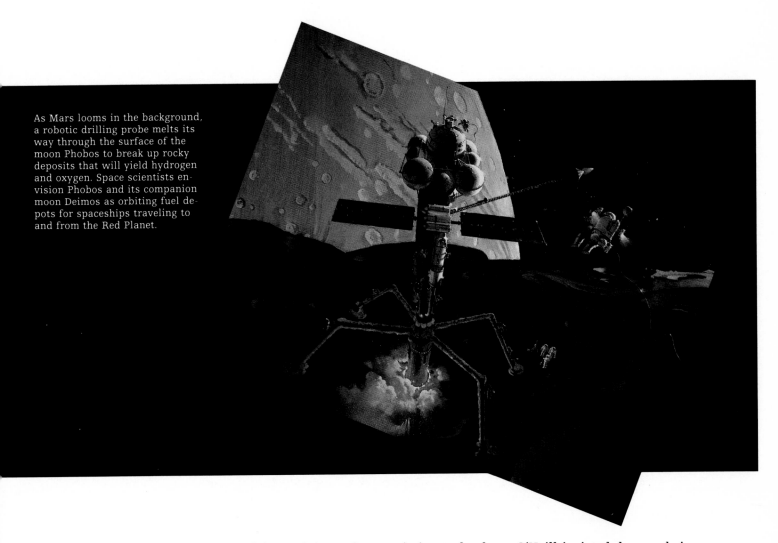

As Mars looms in the background, a robotic drilling probe melts its way through the surface of the moon Phobos to break up rocky deposits that will yield hydrogen and oxygen. Space scientists envision Phobos and its companion moon Deimos as orbiting fuel depots for spaceships traveling to and from the Red Planet.

claim. Relying only on existing technology, O'Neill insisted, human beings could construct vast artificial worlds that might house and feed as many as a million people.

In voicing the idea of free-floating space colonies, O'Neill joined a long line of confabulators, from Jules Verne to rocket scientist Wernher von Braun. J. D. Bernal, for instance, in addition to his notions about inhabiting asteroids, had imagined a spherical habitat about ten miles in diameter that would be built in space. "Owing to the absence of gravitation," Bernal had ventured, "its construction would not be an engineering feat of any magnitude." The sphere would be covered by a thin shell of transparent material, whose "chief functions," Bernal noted, "would be to prevent the escape of gases from the interior, to preserve the rigidity of the structure, and to allow the free access of radiant energy. Immediately underneath this epidermis would be the apparatus for using this energy either in the form of a network of vessels carrying a chlorophyll-like fluid or some purely electrical contrivance. In the latter case the globe would almost certainly be supplied with vast, tenuous, membranous wings which would increase its area of utilization of sunlight." Bernal thus anticipated by half a century the photoelectric solar cell, multitudes of which would figure into NASA's plans for an orbiting space station and make up the wings of many a NASA probe.

Although he drew on the ideas of Bernal and others, O'Neill was the first to approach the issue of space colonies analytically, complete with reams of computations. The initial problem he tackled was where such habitats should be built. In 1969, he put the question to his Physics 103 class at Princeton. Assuming that civilization is destined to expand beyond the bounds of Earth, he asked his students, would the best place to settle be another body in the Solar System—the Moon, say, or Mars? O'Neill had pretty much made up his mind when he posed the question, and his students reached the same conclusion: No moon or planet offered ideal conditions in terms of such critical factors as gravity, atmosphere, availability of sunlight, and length of day. A more hospitable environment could be simulated by artificial worlds orbiting Earth. And those worlds need not be claustrophobic; as Bernal had suggested, there appeared to be no obstacle to building and sustaining a space habitat several miles wide in a low-gravity environment.

O'Neill began penning papers to that effect, but it took four years and numerous rejections at the hands of his scientific peers before one of his articles was accepted by the journal *Physics Today* in 1974. That article, and a conference on space colonies O'Neill convened at Princeton, brought him a sudden spate of media attention. His speculations filled the pages of publications such as the *New York Times, Time,* and *Newsweek.* Television talk-show host Johnny Carson chatted with him, and an article in *People* magazine noted a resemblance between the physicist and the cerebral alien Mr. Spock of the television series "Star Trek." Although O'Neill realized that the public might find the notion hard to accept, he did not hesitate to repeat that the first orbiting space settlements could be set up at a relatively small cost, about equal to the total funding of the Apollo program. Perhaps prodded by the publicity, NASA in 1975 began supporting research into the technological requirements of such an enterprise.

A CONCLAVE OF WOULD-BE COLONIZERS

To launch the effort, the agency sponsored an intensive, ten-week study that summer at the NASA Ames Research Center near Stanford University. Joining O'Neill there were space-habitat enthusiasts of diverse stripes, including astronomers, physicists, architects, and economists. A few of those attending were undergraduates. Eric Drexler, a 20-year-old student at the Massachusetts Institute of Technology (MIT) who had been enthralled since high school with the idea of using asteroids to colonize space, arrived at the seminar with a handful of like-minded wunderkinds from the Cambridge area. "Several of them turned out to know more than most of the faculty members in the study," recalled one of the conferees, planetary scientist T. A. Heppenheimer. "They did a great deal of useful work."

In weighing the various possible designs for a habitat, the conferees paid special attention to two hazards associated with prolonged stays in space.

Beginning in the mid-1970s, Princeton physicist Gerard K. O'Neill—shown here holding a model of a proposed space habitat—championed free-floating colonies as the best way to settle what he called "the high frontier." O'Neill favored a cylindrical design for a space colony because it could be rotated to create artificial gravity and provided the greatest internal land area for a given structural mass.

The first was exposure to potentially deadly radiation, whether in the form of cosmic rays from deep space or x-rays and other emissions from solar flares. The short-term effects of radiation were eerily evident to the first astronauts, who reported seeing random flashes of light while in orbit that were in fact caused by nuclear particles bombarding their retinas. Earth's occupants are protected from the worst radiation hazards by the planet's dense atmosphere and powerful magnetic field. But orbiting colonists would be at constant risk and would require a permanent shield.

The second hazard was perhaps less dire but equally disturbing—the subtle degenerative effect of prolonged weightlessness on the human body. Without the constant resistance imposed by gravity on Earth, the heart and other muscles shrink, blood vessels constrict, fluid levels decrease, and bone wastes. In just one month, for example, the heel bone can lose five percent of its mass. Rigorous exercise can slow the rate of deterioration, and the body might conceivably adjust to weightlessness over a period of years. But any degree of enfeeblement would cast a pall over life in a space colony. Weighing the health concerns, the planners gathered for the summer study concluded that a vigorous Earth-like existence on the colony required a gravitational force comparable to that on the home planet.

Gravity could be simulated by causing a habitat to rotate, producing a centrifugal force that would enable colonists to plant their feet on the outer wall of the structure and walk about with their heads pointed toward the axis of the revolving colony. Yet the idea of relying on centrifugal force to lend the inhabitants weight posed some physiological problems of its own. Occupants of a rapidly revolving spacecraft would be bothered by the Coriolis effect: the sideways deflection of moving objects and fluids relative to a fixed point on a spinning surface. Thus the sensitive fluids in the inner ear, responsible for the sense of balance, would be deflected in the direction of the spin even when the spacefarer was standing still, producing a feeling of wooziness. One of the undergraduates attending the summer study, a premedical student named Larry Winkler, calculated that a spin rate of more than one revolution per minute was likely to cause motion sickness. His objection spelled trouble for a modest-size cylindrical habitat advanced by O'Neill: At a mere 600 or so feet in diameter, O'Neill's structure would have to rotate three times a minute to simulate normal Earth gravity. And the nausea experienced by the colonists in such a fast-revolving space habitat would not be the only manifestation of the Coriolis effect. As Heppenheimer put it, "A colonist might be watering his lawn from a hose and see the stream curve over onto his neighbor's barbecue. Or he might be taking a shower and the spray from the showerhead would curve onto the floor."

Concern over the spin rate led the conferees to think in terms of a larger structure that could achieve Earth-like gravity at a more comfortable pace. Because centrifugal force increases with distance from the spinning object's axis of rotation, a wide object revolving at a slower rate than a narrow one can generate the same force at the periphery. One design that was considered was a giant sphere similar to that envisioned by Bernal. However, residents within such a rotating bubble would notice a pronounced and unsettling drop in gravity as they moved from the equator, with its wide radius, toward the poles. And the disorienting effect would be amplified by the topsy-turvy layout of a rounded world: Inhabitants would look up to see houses hanging from the far side of the sphere. O'Neill's plan for a more cylindrical structure *(page 65)* avoided some of the pitfalls of Bernal's design, with its polar extremes. But other participants in the summer study wanted to go further. In a few evenings of feverish work, they worked out an option that would, in effect, put all the colonists at the equator. The design that emerged was a large rotating tube known as the Stanford torus *(page 66),* and it was ultimately endorsed by the study group as a whole.

REINVENTING THE "LIVING WHEEL"
The potential advantages of the torus were not unknown to space-habitat enthusiasts. As early as 1929, an Austrian engineer who went by the pen name of Hermann Noordung had envisioned a space station he called a *Wohnrad,*

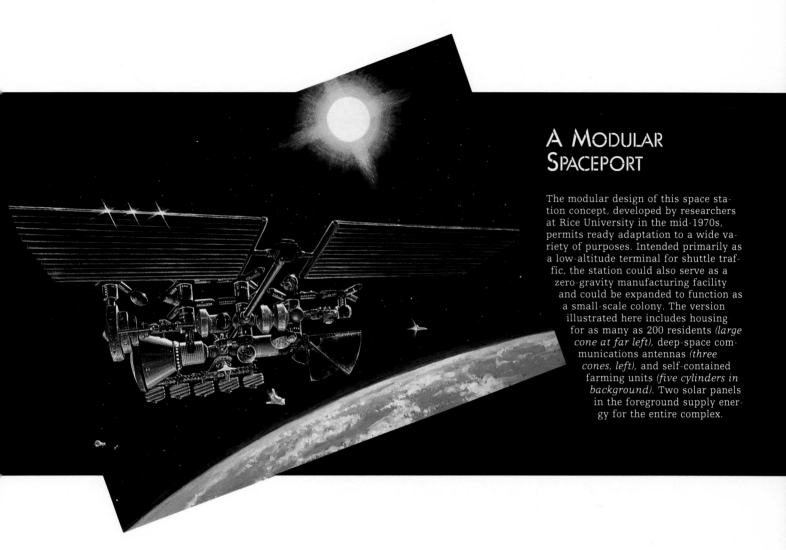

A Modular Spaceport

The modular design of this space station concept, developed by researchers at Rice University in the mid-1970s, permits ready adaptation to a wide variety of purposes. Intended primarily as a low-altitude terminal for shuttle traffic, the station could also serve as a zero-gravity manufacturing facility and could be expanded to function as a small-scale colony. The version illustrated here includes housing for as many as 200 residents *(large cone at far left)*, deep-space communications antennas *(three cones, left)*, and self-contained farming units *(five cylinders in background)*. Two solar panels in the foreground supply energy for the entire complex.

or "living wheel," that would rotate to provide artificial gravity and would derive its power from the Sun. Twenty years later, rocket scientist Wernher von Braun proposed a more sophisticated torus plan that would include a system of ballast tanks to keep the structure revolving at a steady rate even as the inhabitants moved about and shifted the center of mass. The configuration was later popularized in the film version of Arthur C. Clarke's fantasy *2001*, which featured a pinwheel-like space station pirouetting gracefully to the strains of a waltz.

What the summer-study participants came up with was the first comprehensive torus design that addressed the problems raised by actual spaceflights, including the need for a radiation shield. Their proposal called for a free-floating barrier ringing the torus proper that would be packed with lunar soil, presumably the slag left over from mining operations on the Moon. On the inner rim of the torus—the side closer to the hub—mirrored panels overlapping like the slats on venetian blinds would allow sunlight reflected by a large disk-shaped mirror positioned above the colony to filter through to the interior; this arrangement would screen out cosmic rays, which do not reflect, and would keep harmful solar radiation to a minimum. Slightly more than a

Life and Work in a Hollow Sphere

In 1976, physicist Gerard O'Neill, along with NASA scientists, conceived a design for a space colony *(right)* whose main feature is a hollow sphere measuring nearly one mile in circumference. Rotating at sufficient speed to simulate Earth's gravity, the sphere would house up to 10,000 people on its inner surface. A ring of reflecting panels and two large mirrors at either end of the sphere direct sunlight to the interior. Crops would be grown in two sets of tubelike coils beyond the large mirrors. At each end of the central axis are docking and manufacturing areas, as well as radiator panels for dispersing excess heat.

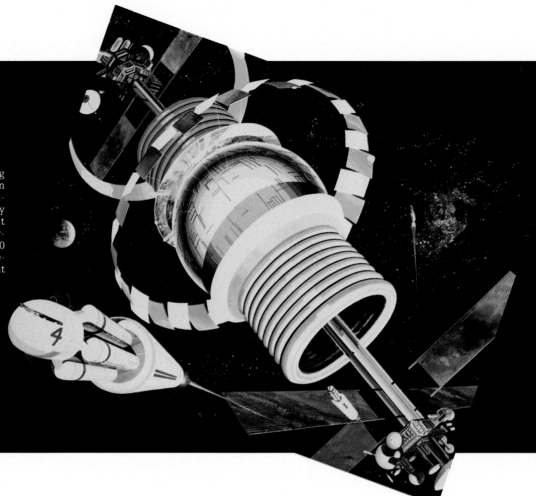

mile in diameter, the torus would provide 1g (Earth gravity) at about one revolution per minute. The central hub—linked to the torus by thick spokes housing elevators—would provide weightless conditions for docking procedures, construction projects, and recreation.

Among its other advantages, the torus would have a spacious feel to it. The outer rim of the tube would serve as the "floor" of the habitat, while the inner rim, with its panels admitting light, would be the ceiling. Gazing down the length of the torus, colonists would behold a landscaped world curving gently upward toward the horizon and graced with modular houses and lawns rising on terraces from a central arcade. As the summer-study participants put it in their report, the habitat would be "a busy community without skyscrapers and freeways; a city which does not dwarf its inhabitants. The human scale of the architecture is emphasized by the long lines of sight, the frequent clusters of small fruit trees and parks, and the sense of openness produced by the broad expanse of yellow sunlight streaming down from far overhead."

To provide a sense of ecological diversity, residential areas would alternate with agricultural zones, so that a short walk in either direction would bring colonists to verdant tiers of pastures, gardens, and ponds. In a closed cycle

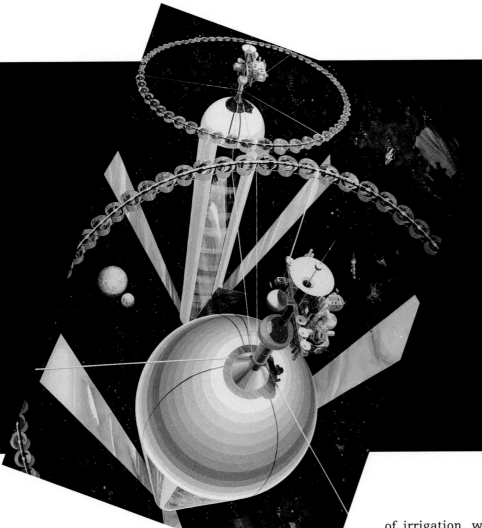

Cylinders for Island Three

The illustration at left shows two identical versions of an alternative O'Neill concept, dubbed Island Three. Each cylinder is four miles in diameter and twenty miles long, and would accommodate several million inhabitants. Three gigantic mirrors running the length of each cylinder are hinged at one end so that they could be opened and closed, simulating day and night for the residents within, who would occupy three strips of landscape between the mirror panels. Each colony also includes a ring of smaller cylinders set aside for agriculture, allowing independent, customized control of lighting and climate.

of irrigation, water would spill down from the ponds—set at the top level and stocked with fish—to nourish crops on the next tier; staples might include corn, soybeans, and alfalfa. The runoff from the gardens would in turn be channeled to animal pens at floor level. (The study estimated that the 10,000 or so colonists could maintain 60,000 chickens, 30,000 rabbits, and sizable herds of cattle.) The water would then pass through a purifying plant below the floor before being shunted through the loop again. The process would provide enough food to maintain each inhabitant on a healthful mixed diet of about 2,400 calories a day.

The colony would be self-sustaining in other respects as well. Energy would be derived exclusively from the Sun, and the air would be purified and waste products recycled without the natural filters provided on Earth by the oceans and the atmosphere. The abundant vegetation would help significantly by removing carbon dioxide from the habitat's air and releasing both oxygen and water vapor. But colonists would have to rely on advanced technologies for the rest. Dehumidifiers concentrated in the agricultural areas would condense excess moisture to prevent saturation of the air and to replenish the

SPACE WHEELS AND MIRRORS

A 1975 study directed jointly by Stanford University and the NASA Ames Research Center came up with the design illustrated at right: an immense, rotating, wheel-shaped structure more than a mile in diameter, accompanied by a large, free-floating polished disk. Pitched at a 45-degree angle, the disk would reflect sunlight toward an inner ring of mirrors, which would in turn direct the light into the wheel's tubular outer ring. Living and working space for about 10,000 inhabitants would be organized within the tube along its outer circumference. Because of the disk's tilt, opposite halves of the tube would experience day alternating with night as the wheel rotated.

supply of drinking water. And the freshwater reserves would be further bolstered through a complex process known as wet oxidation that would purify the polluted agricultural runoff and the residential sewage by subjecting the wastes to pressure and heat. The carbon dioxide given off by wet oxidation would be channeled to the gardens to promote plant growth, while solids in the effluent would be filtered out and processed to produce feed for the animals and fertilizer.

Among the few materials that would have to be imported to the colony would be those required for commercial activity in the hub, including satellite production and repair, and the manufacture of components for additional habitats. Access to the Moon and its various minerals could be facilitated by locating the colony at a nearby libration point, one of five positions between the Moon and Earth where the gravitational attraction of the two bodies is balanced, so that a spacecraft in the vicinity will maintain a stable position in relation to both. Alternatively, a captured asteroid could be guided into orbit near the colony, in which case its residents might become independent of outside sources industrially as well as agriculturally.

Overall, the summer study's vision of life in the torus offered a tantalizing glimpse of a new space-based ecology in which virtually nothing would go to waste. The conferees conceded that questions remained to be resolved before the dream could become a reality. But the very fact that most of the colony's anticipated needs could be met through existing or emerging technologies held out the promise that artificial worlds would become permanent fixtures in space in the twenty-first century.

An important spur for colonization might be dwindling energy resources on Earth and the need to harness the Sun's power in its purest form, above the blanket of Earth's atmosphere. Gerard O'Neill, in his influential book *The High Frontier,* published in 1977, maintained that the economic necessity of obtaining cheap fuel for Earth would force humans off their home planet to build and launch solar collectors that could beam the energy back to the planet. The establishment of permanent space settlements would not be the result of some utopian impulse but, as he put it, "the logical consequence of high-level industrial activity in space."

OBSTACLES IN THE PATH

O'Neill's writings and the report of the NASA summer study prompted scientists to look carefully and sometimes critically at the proposed solutions to the many problems of sustaining life in space. For example, the reliance on artificial gravity, even at the slow rate of rotation allowed by large structures, struck some analysts as potentially troublesome. Theoretically, an orbiting habitat, once set in motion, would continue to rotate steadily at that pace for lack of atmospheric resistance. But as Wernher von Braun recognized in his early torus design, shifts of weight within the structure would affect the rate of rotation, with disorienting effects on the occupants. Conceivably, twenty-first-century engineers might come up with an energy-efficient way of maintaining steady rotation, but the complexity of the problem—coupled with concern that the Coriolis effect might still disturb colonists at rates as low as one revolution per minute—led researchers to seek alternatives to round-the-clock artificial gravity.

One innovative option, designed and tested in the late 1980s by Peter Diamandis at the Massachusetts Institute of Technology's Man-Vehicle Laboratory, is the so-called Artificial Gravity Sleeper: a rotating bed in which the subject sleeps with his or her head at the axis to produce centrifugal force toward the feet with no feeling of dizziness or nausea so long as the eyes are closed. The device, known informally as Robocot, would allow spacefarers in a nonrotating craft to supplement their regular program of exercise with a healthy dose of gravity nightly—a recourse that might well be more practical for small crews on long journeys of exploration than for orbiting colonies of thousands intent on conserving power.

In any case, some form of artificial gravity could be essential for vegetation as well as people. Soviet cosmonauts on long-term missions in the Salyut space stations found that the weightless environment impaired the ability of plants to distribute their fluids; seedlings fared better when they were rotated in a centrifuge—the horticultural equivalent of a Robocot. Assuming that future space habitats can do without such machines by maintaining a stable rotation rate, their agricultural areas may still look more like laboratories than like gardens on Earth. Researchers at NASA's Biomass Production Center at Cape Canaveral, Florida, have been testing an intensive form of hydroponics—soilless plant growth—in which crops such as wheat and soybeans sprout from a porous membrane inside long tubes filled with water and nutrients. The technique promises a higher yield and a more reliable growth cycle than agriculture based on soil, which might eventually be exhausted by the nutritional demands of a large colony.

In the end, the greatest challenges posed by colonization may be not biological but social. Even the largest habitats would at first seem alien and confining to settlers accustomed to changing vistas and unfettered movement. And the close atmosphere of the colony could turn minor complaints and conflicts among settlers into serious disputes that would threaten the very existence of the community. Conceivably, the need for tight social control could foster rigid discipline within some habitats. But order might also be maintained through cooperative effort, assuming the members were united by some higher purpose. The settlers in one habitat might be ecological utopians, for example, dismayed by the destruction of Earth's natural resources and determined to maintain a pristine environment in space. Those in another might be members of a persecuted sect, pilgrims ready to brave life on a strange frontier in order to regain their spiritual freedom. Even colonies whose objectives were strictly economic or scientific might well evolve solemn rituals to inspire members with a sense of collective purpose.

Although orbiting habitats would maintain contact with Earth and with one another, the members of each community would doubtless pride themselves on their independence, and some groups might grow resistant to outside influence in the manner of pioneer colonies on Earth. Ultimately, one or more of these rebellious societies—whose founders had passed their strict habits of self-reliance on to a second generation of settlers with no nostalgic bonds to the planet—would be prepared to take the next fateful step. Severing the thin gravitational bond holding them to Earth, they would journey starward to occupy a higher frontier.

CULTIVATING THE RED PLANET

One objective of outward-bound colonists could well be Mars. By the time sizable space habitats are in place around Earth, select teams of astronauts may have completed the construction of permanent stations on Mars. By one estimate, such bases could be ready for occupation within a few decades of the first Martian landing.

The thin atmosphere will provide little protection against dangerous radiation, so living areas will have to be shielded by soil, but plants will be exposed to direct sunlight through plastic panels treated to screen out the harmful ultraviolet rays. Because the force of gravity on Mars is roughly

CREATING EARTH ON MARS

Of all the celestial bodies humanity has been able to examine so far, Mars offers the best prospect for a planetary home away from home. The assessment might seem unwarranted, given the Red Planet's present condition *(below)*. A frigid desert world whose lack of a protective atmosphere exposes the surface to lethal ultraviolet radiation from the Sun, it bears very little resemblance to the nurturing blue Earth, and any would-be Martians must plan on dwelling there in habitats that shield them from the grim environment.

But a few visionary scientists dream of a less restricted life for earthly colonists. Through a process known as terraforming, they would convert the barren globe into a self-sustaining ecosystem, protected and warmed by its own atmosphere, flowing with water, and populated by living organisms that rival Earth's in their number and variety. Some of Mars's topographic features *(pages 70-71)* suggest that eons ago—about the time life first emerged on Earth—conditions on the two planets might have been remarkably similar. Terraforming advocates believe that Mars could return to that hospitable state—although the requisite planetary engineering would take millennia.

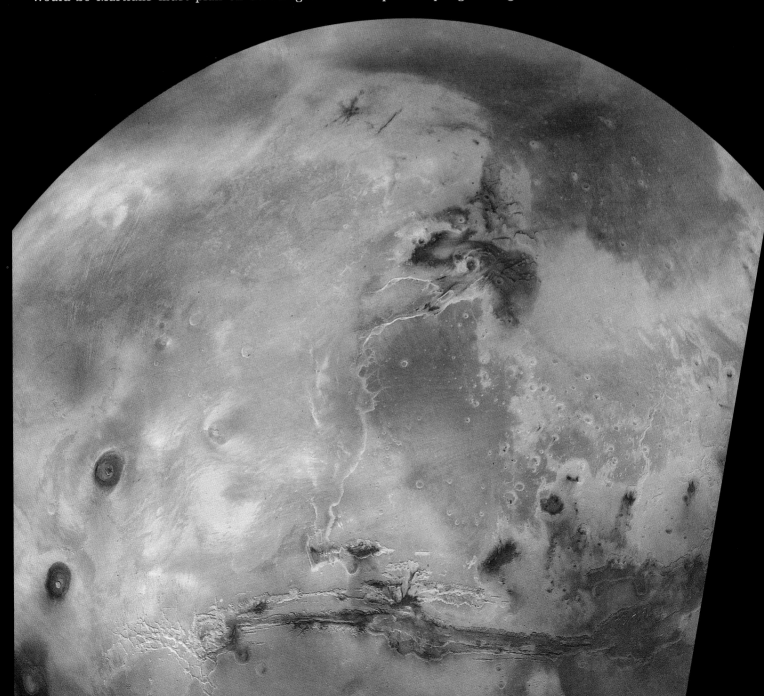

Data from infrared detectors on the Viking orbiters revealed that the permanent icecap covering Mars's north pole, shown at left in a photomosaic of about 400 images taken by the *Viking 2* orbiter, is pure water ice. Some terraforming schemes suggest darkening the icecap with algae to enhance heat retention, thereby shrinking the cap and melting the permafrost—releasing vital water that would also contribute to a thickening atmosphere.

CLUES TO AN EARTH-LIKE PAST

Close-ups of the Martian surface transmitted by the twin probes *Viking 1* and *Viking 2* in 1976 revealed a history of geologic activity strikingly familiar to terrestrial scientists. Enormous volcanoes *(right)* once spewed forth torrents of lava, and vast quantities of liquid water must have carved the network of dry washes and riverbeds *(far right)* that weave across the now desiccated landscape.

Many scientists conclude that the young Mars resembled its sister Earth and was much warmer and enveloped by a thicker atmosphere than it possesses today. Some go further, postulating that the Red Planet—with its ration of water, nitrogen, and carbon dioxide—may have been able to spawn life of its own during its first billion years.

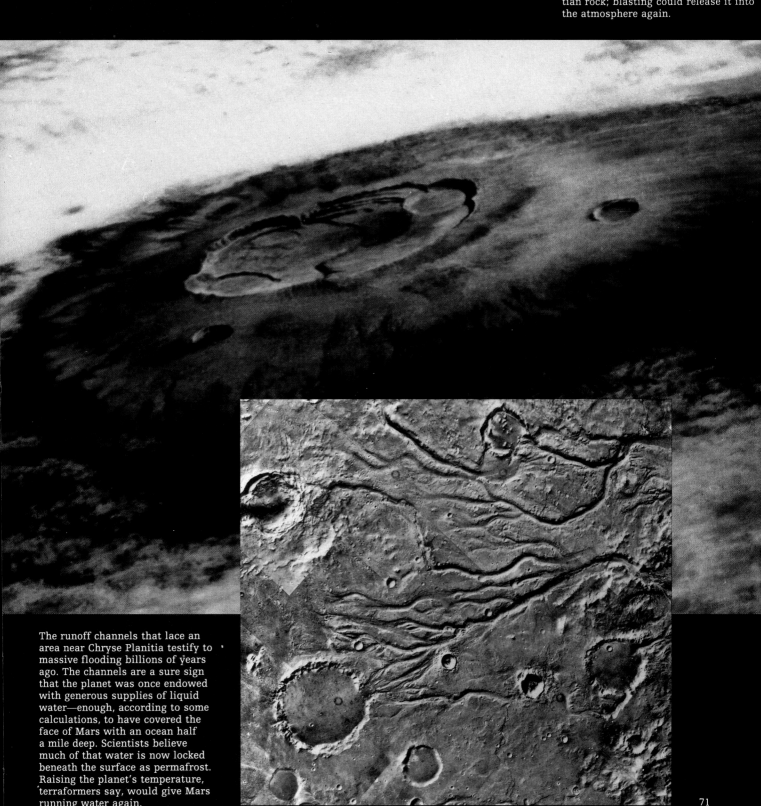

The largest known volcano in the Solar System, Olympus Mons rises some 16.8 miles high, three times the elevation of Mount Everest; were it set down in Europe, it would cover most of France. Martian volcanoes once spewed volumes of carbon dioxide, creating a dense atmosphere that trapped solar heat to warm the planet's surface. Most of the carbon dioxide is now locked up in Martian rock; blasting could release it into the atmosphere again.

The runoff channels that lace an area near Chryse Planitia testify to massive flooding billions of years ago. The channels are a sure sign that the planet was once endowed with generous supplies of liquid water—enough, according to some calculations, to have covered the face of Mars with an ocean half a mile deep. Scientists believe much of that water is now locked beneath the surface as permafrost. Raising the planet's temperature, terraformers say, would give Mars running water again.

The dusty plains and dry canyons of today are transformed into oceans, rivers, and continents in this artist's rendition of a terraformed Mars of the distant future. Safeguarded by a newly created atmosphere, the Red Planet could become another garden world of the Solar System —and a cradle of new civilizations.

one-third that on Earth, it could turn out to be adequate for the growth of plants and the health of humans. But the Martian day-night cycle (approximately the same as on Earth) will provide stations there with a smaller fund of solar energy than habitats in Earth orbit will enjoy. To compensate, Martian pioneers might derive fuel for heating and the operation of vehicles by converting the abundant carbon dioxide in the planet's atmosphere into carbon monoxide and oxygen, gases that could be burned together in combustion engines; the conversion process, however, would itself require a significant input of solar or electrical energy. One peculiar peril facing residents on Mars would be extremes of weather. Global dust storms raised by gale-force winds would periodically obscure the Sun for weeks on end and confine colonists to their shelters like troops under siege.

Such rigors and the sheer distance of Mars from Earth will probably restrict visits to the Red Planet for some time to small groups of scientists and technicians, living under conditions reminiscent of those at the South Pole in winter. But the trickle of travelers to Mars could eventually become a flood if the Martian environment could somehow be made more Earth-like—a hypothetical process known as terraforming. The term was coined in the 1940s by science-fiction writer Jack Williamson, who composed a tale about engineers who cloaked "in green life all the riven stone of a world born dead." This compelling idea of grafting biology onto a planet possessed of mere geology was echoed by a serious student of the cosmos, Fritz Zwicky, the eccentrically brilliant California Institute of Technology astronomer who discovered supernovae, or exploding stars. In 1948, Zwicky suggested that Mars and even the planets beyond might be made hospitable to life if they could somehow be nudged closer to the Sun. A few years later, Arthur C. Clarke offered an innovative amendment to that proposal in his novel *Sands of Mars.* Clarke's scenario avoided the seemingly impossible task of relocating Mars, but it was still purely conjectural: He envisioned initiating a slow nuclear-fusion reaction on one of the two small Martian moons that would transform that moon into a miniature sun, thus melting the water and carbon-dioxide ice on the planet and fostering the gradual development of a dense life-sustaining atmosphere.

PLANETARY ENGINEERING
By the 1970s, the prospect of terraforming Mars and other bodies in the Solar System had spawned a new speculative discipline—planetary engineering. One prominent enthusiast, astronomer Carl Sagan, proposed in 1973 that Mars could be warmed significantly simply by darkening its highly reflective icecaps to improve heat retention; this might be accomplished either by showering the polar regions with black dust (a pulverized asteroid might do the trick) or by introducing dark algae that would spread like a stain across the Martian poles. Terraforming won official recognition of sorts when some of the scientists gathered for the space-colony summer study at the NASA Ames Research Center in 1975 examined various scenarios for making Mars

habitable by increasing its temperature, water, and oxygen levels. The scientists concluded that there was "no fundamental, insuperable limitation to the ability of Mars to support terrestrial life."

Interest in revamping the Martian environment was not limited to the United States. Among those drawn to the topic was the British atmospheric chemist James Lovelock, best known for his advocacy of the Gaia hypothesis, which holds that Earth constitutes a self-regulating system in which life forms create conditions conducive to their maintenance. As it happens, this Earth-centered theory grew out of attempts by Lovelock and others to gauge the prospects for life on Mars experimentally.

One way to approach the problem was to analyze the Martian atmosphere and compare it with Earth's. The approach seemed straightforward enough, but the selection of Earth's atmosphere as a standard raised some provocative questions about the chemical causes and effects of life. What conclusions could be drawn, for example, from the abundance of oxygen on Earth—a full 21 percent of the atmospheric mix? Such a high quotient could scarcely be a precondition for life, since free oxygen reacts readily with other elements and would quickly be reduced to a trace in the atmosphere unless replenished by existing plants and algae. Evidently, Earth's oxygen supply increased grad-

Warming up Mars. Pondering the cold, dry, almost airless conditions of Mars, British theorist James Lovelock *(far left)* imagined making it hospitable to human habitation by introducing gases that would trap heat, creating a warming greenhouse effect. American researcher Christopher McKay *(center)* calculated that several million tons of the gases might be needed each year for maintenance. British researcher Martyn Fogg suggested that a quicker application of heat might be required—ten million thermonuclear bombs to liberate subsurface carbon dioxide and water.

ually along with its emerging flora. But what kept the process from spiraling out of control? If the oxygen quotient increased by only a few percentage points, Lovelock estimated, Earth's atmosphere would become so volatile that fires would rage unchecked and consume all living things. He concluded that there were natural mechanisms at work that supplied enough oxygen to support vigorous activity by animals as well as plants without exposing them all to catastrophic oxidation. For example, a surge in plant growth brought on by a warming climate would increase the production of oxygen, but it would also deplete carbon dioxide, a so-called greenhouse gas that traps heat in the lower levels of the atmosphere. As the blanket of carbon dioxide grew thinner, the planet would gradually cool and plant growth would subside, holding the oxygen supply within safe limits.

TURNING UP THE HEAT

For Lovelock, the notion that Earth regulates its processes to sustain life held out hope for efforts to terraform Mars. To be sure, the gaseous makeup of that planet—with 95 percent carbon dioxide and only a trace of oxygen—presented a stark contrast to Earth's atmospheric blend. But plants could thrive in a rich mix of carbon dioxide, and Lovelock suspected that once the seeds of life were planted on Mars, natural processes would gradually enrich the atmosphere until it reached a stable state that would support a complex ecosystem. The real stumbling block was not that the existing Martian atmosphere was poor in oxygen but that it was so thin and thus failed to retain the heat and moisture essential to abundant life.

Like others who had pondered the problem of cultivating Mars, Lovelock agreed that the first essential task was to raise the temperature of the planet in order to unlock its frozen reserves of water and carbon dioxide, whose evaporation would then enrich the atmosphere. But he saw no need to wait indefinitely for scientific breakthroughs that would allow Mars to be propelled closer to the Sun or heated by an orbiting fusion furnace. Instead, he put forward an ingenious scenario involving existing technologies. Detailed in *The Greening of Mars*, a science-fiction novel set in the near future that Lovelock coauthored with Michael Allaby in 1984, the scheme called for liberal use of the notorious chemicals known as chlorofluorocarbons (CFCs), used in spray cans, refrigerators, and air conditioners and implicated by the time the book was published in the depletion of Earth's protective ozone layer. On oxygen-poor Mars, Lovelock and Allaby noted, there was no ozone layer to endanger. Introduced into the atmosphere there, the CFCs would instead create a benign greenhouse effect, thanks to their ability to trap the Sun's energy a thousand times more efficiently than even carbon dioxide.

In *The Greening of Mars,* the suspect chlorofluorocarbons, having been outlawed on Earth, are collected and dispatched to Mars in thousands of rockets that explode on impact, casting a heat-catching chemical shroud over the planet. Some of the same rockets that disperse the gases also seed the Martian polar regions with the spores of dark algae known to thrive in the snows of Antarctica; the organisms quickly proliferate and further promote heat retention. In time, the conspiring effects of the algae and the CFCs cause the polar icecaps to shrink and the permafrost to melt, creating lakes, rivers, and marshes whose mists and vapors thicken the atmosphere.

Within two decades of the missile bombardment, the first settlers arrive and establish colonies along the nearly temperate equator. The planet's air pressure is still low and oxygen sparse, so the pioneers must at first live in enclosed habitats and venture forth only in pressurized suits. Nonetheless, they can begin to cultivate the soil, tilling fields that are warmed by orbiting reflectors made from shiny solar sails that helped propel the colonists to their new home. The emergence of green plants further conditions the atmosphere. After a few decades, the air pressure on Mars is comparable to that at the top of the highest mountains on Earth, and settlers can move about equipped only with oxygen masks. Ultimately, the continued greening of the Red Planet—and the constant release into the air of oxygen derived by separating Martian ores from their oxides or by simply warming the soil enough to liberate the oxygen bound up there—will enable descendants of the first pioneers to breathe freely outdoors.

This compelling scenario has drawn a variety of responses from exponents of planetary engineering. In the opinion of some analysts, the idea of somehow creating a Martian greenhouse effect to promote terrestrial-type life warrants serious consideration. Christopher McKay, a research scientist at NASA Ames Research Center who has studied the atmospheric and geological resources of Mars with an eye toward future colonization, contends that if greenhouse gases could be disseminated in a systematic way—by technicians at Martian bases, say, releasing CFCs derived from indigenous materials—significant heating might eventually result.

A cautionary note, however, has been sounded by Martyn Fogg, a contributing editor of the *Journal of the British Interplanetary Society.* Fogg doubts that global warming alone would contribute significantly to the greening of Mars. Citing data suggesting that most of the carbon dioxide on Mars is locked up not in ice but in rocks, he argues that even a substantial rise in temperature would do little to thicken the atmosphere. Indeed, the gas might have to be blasted out in a series of thermonuclear explosions. And even once the Martian atmosphere is dense with carbon dioxide, he adds, it will take at least a thousand years of natural photosynthesis, supplemented by mechanical production of oxygen, to yield breathable air. "Mars will only be said to have been fully terraformed after millennia of continuous planetary engineering," Fogg

In a scene scheduled to become fact in late 1995, the *Galileo* space probe—launched in October 1989—orbits Jupiter in the vicinity of the moon Io *(far left)*. Io is among the most intriguing of the Solar System's outer worlds, spewing sulfur hundreds of miles above its rocky surface from volcanoes powered by tidal interactions with the parent planet.

concludes. "The task will not be one which can be set in motion and then left to run its course but will need continuous management by a technologically advanced civilization."

TERRAFORMING IN THE TWILIGHT ZONE

If schemes for cultivating Mars sometimes blur the line between science and fiction, the distinction is all but obliterated by proposals to terraform bodies in the outer Solar System. The four moons of Jupiter discovered by Galileo, for example, have inspired more than a few imaginative leaps by planetary engineers. The largest member of that quartet, Ganymede, has twice the mass of Earth's moon, endowing it with enough gravity to hold an atmosphere; and its surface is icy, raising fertile possibilities if sufficient heat could be applied. Yet Ganymede and its lunar neighbors are roughly three times as far from the Sun as frigid Mars, and the solar energy available to the satellites is discouragingly slight.

Martyn Fogg has suggested a way around the problem in a scenario that takes terraforming to dizzying heights. Fogg argues that the heat needed to thaw out the frozen Galilean moons could be derived from Jupiter itself by transforming that huge gaseous planet into a star. The energy required to convert the planet would come from a black hole: a superdense object that exerts such intense gravity even light cannot escape from its hold. Astronomers speculate that a tiny, primordial black hole (PBH)—a theoretical entity that formed in the earliest moments of the universe—may be lurking close to, or even within, the Solar System. If such a phantom could be located and drawn into a decaying orbit around Jupiter, perhaps by using a captured asteroid as a kind of "gravitational tug," the PBH would eventually descend into the planet's core. Once lodged there, the ancient black hole would make the most of the opportunity and begin sucking matter into its insatiable maw, emitting energy that would kindle surrounding layers of gas within Jupiter and warm the Galilean moons.

The process would unfold so slowly that, by Fogg's estimate, the Galilean moons would be habitable for 12 to 60 million years, ample time for civilizations upon them to rise and fall. After another 440 million years or so, the black hole would have finished its work and Jupiter would no longer exist. The entire scenario, Fogg freely admits, presses up against "the ultimate boundaries of the possible."

CHARTING A PATH TO THE STARS

Long before Martian colonists walk the planet unimpeded by gas masks or settlers on Ganymede bask in the glow of an incandescent Jupiter, trailblazing spacefarers will venture beyond Mars and negotiate the asteroid belt to explore—and perhaps exploit—the solid bodies of the outer Solar System. Given the length and complexity of such missions, potential landing sites will be carefully selected on the basis of data furnished by robotic probes. The spacecraft *Galileo,* for example, launched in 1989, is scheduled to reach Jupiter in 1995 and spend two years surveying its moons, an effort that should aid mission planners of the next century seeking the best target for a human expedition beyond Mars. Once an objective is defined, a more versatile probe, equipped to land at the site, will be dispatched to analyze the surface and any surrounding atmosphere for resources that would be of use to a crew establishing a base there.

Bridging the huge void that lies between Mars and Jupiter—and the even greater planetary intervals waiting beyond—would require enormous expenditures of time and fuel if future spacefarers had to rely on present methods of propulsion. Fortunately, breakthroughs are in sight. Among the scientists laying the theoretical groundwork for trans-Martian flights is Robert M. Zubrin, a senior engineer with the astronautics division of the Martin Marietta Corporation. Zubrin began musing about interplanetary propulsion systems during his days as an undergraduate mathematics major at the University of Rochester, New York. Even before receiving his doctorate in nuclear engineering from the University of Washington, he coauthored a seminal paper with Dana Andrews of Boeing Aerospace, laying out the details of an entirely new form of transport, the magnetic sail *(opposite).* Unlike a

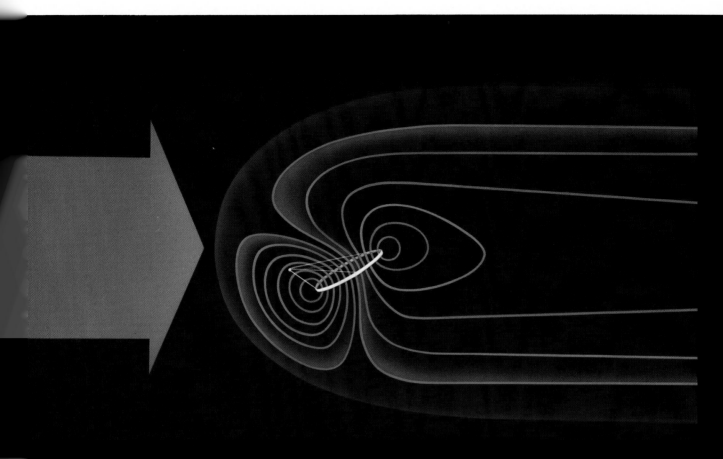

MAGNETIC PROPULSION

Future spacefarers may cross the interplanetary gulfs by magsail—an elegantly simple propulsion system proposed by American aeronautical engineers Robert Zubrin *(near right)* and Dana Andrews. A magsail's key component—still hypothetical—is a long loop of superconducting cable *(white in the diagram above)* linked to a ship *(white dot)* by a rigging of shrouds. When electrical current is sent through the cable, it flows continuously, creating a sustained magnetic field *(blue)*. This field deflects the solar wind *(arrow)*, a flood of charged particles ceaselessly streaming from the Sun at a million miles per hour. The particles impart much of their momentum to the loop, which tows the ship along. Tilting the magsail permits maneuvering, and switching off the current shuts the system down. In addition to its prospective virtues as a versatile, inexpensive mode of transport, a magsail would help protect spacefarers from dangerous radiation by its shield of magnetism.

solar sail, which would rely on a huge expanse of ultrathin fabric to catch the thrust of particles streaming from the Sun, the so-called magsail would consist of a loop of cable up to twenty miles in diameter carrying an electrical current so powerful that it would induce a sweeping magnetic field around the spacecraft. When the solar wind encountered this artificial magnetosphere, the shock would generate fifty times as much thrust as a solar sail propelling a craft of the same weight.

Tests of the magsail concept will have to await the appearance of superconducting cable, which would eliminate resistance to the electrical current and make it possible to maintain the magnetic field indefinitely. Pending that development, simpler proposals by Zubrin and his colleagues may help to boost the range of spaceflight substantially. As a contributor to NASA's Manned Mars Mission Study, Zubrin has proposed that the agency revive and modify a design it nearly brought to fruition in the early 1970s—the Nuclear Engine for Rocket Vehicle Applications (NERVA).

A SUPPLY OF INDIGENOUS FUEL

NERVA's reactor was designed to heat liquid hydrogen and propel the exhaust through a nozzle at a rate that would provide twice the thrust of conventional rockets. But lifting enough liquid hydrogen off Earth and into space to bring NERVA-powered craft back from Mars or propel them beyond that planet would strain the resources of even the most liberally funded space program. Zubrin's solution is to fuel the nuclear engine with other liquefied gases such as carbon dioxide, the principal component of the Martian atmosphere. Simply compressing the carbon dioxide to six or seven times Earth's atmospheric pressure, a relatively painless chore on Mars given the already frigid temperatures there, would cause the carbon dioxide gas to liquefy. Although the resulting fuel would provide only about one-third the exhaust velocity afforded by liquid hydrogen, that would be more than enough to allow the spacecraft to hop from one spot to another on Mars or to escape the planet altogether. And compressing the carbon dioxide would require far less energy than breaking the gas down into components that could be used by conventional rockets. The nuclear engine could also be adapted to produce a jet of steam from water, a fuel offering somewhat higher exhaust velocity than liquid carbon dioxide and readily available both on Mars and on many moons in the outer Solar System.

The development of an indigenous-fuel rocket would greatly simplify the logistics of planetary exploration. The traditional Apollo-style mission, requiring a landing craft that separates from the mother ship to conserve fuel for the return voyage, would become a thing of the past. Instead, the entire spacecraft—a compact vessel weighing considerably less than mother ship and landing craft combined—would touch down at the selected site and compress fuel for the next leg of the trip.

Such a craft could easily jump from one intriguing target to another, executing a grand tour of a regular satellite system such as Jupiter's. The expedition might begin with a few orbital swings around Io, the innermost member of the Galilean quartet. This would afford the astronauts a close view of volcanic activity on that sulfurous moon, which is heated by tidal interactions with Jupiter and the other satellites. A brief boost and a few subsequent course corrections would then carry the crew to a landing on the smooth icy surface of neighboring Europa—a poor place to settle, perhaps, but a good fueling station for the spacecraft's versatile engine. While stoking the reactor, the astronauts might probe the waters beneath Europa's frozen shell to investigate the possibility that there are life forms lurking in the depths. The expedition would conclude with reconnaissance stops at the two outer Galilean giants, Ganymede and Callisto, whose extensive areas of dark ice may harbor substances of use to occupants of future bases there.

An adaptable nuclear-powered rocket might bring even Saturn and its big brood of satellites within the scope of a round-trip mission. A prime target of investigation would be Titan, Saturn's largest moon and such a rich storehouse of organic compounds that astronomers liken it chemically to a primordial Earth. Zubrin estimates that a nuclear indigenous-fuel Titan explorer (NIFTE), burning hydrogen on the outward journey, could reach that moon from low Earth orbit in four years—much less time than it would take a ship

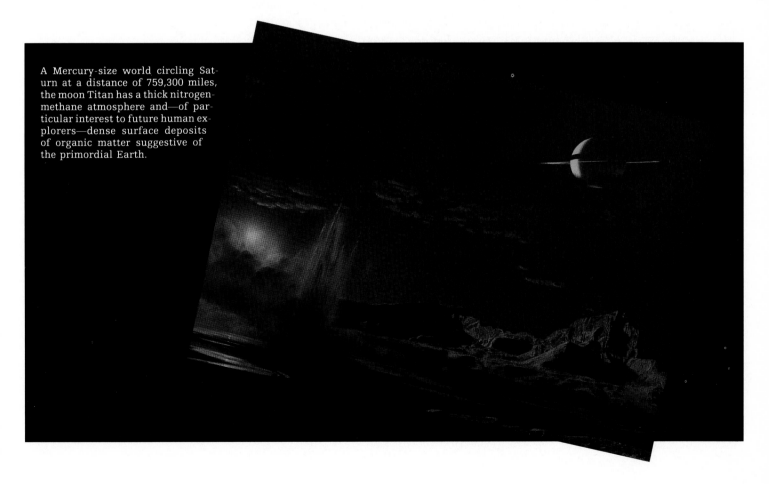

A Mercury-size world circling Saturn at a distance of 759,300 miles, the moon Titan has a thick nitrogen-methane atmosphere and—of particular interest to future human explorers—dense surface deposits of organic matter suggestive of the primordial Earth.

weighed down with fuel for the return journey. The crew would presumably be chosen from a growing pool of astronauts conditioned to long periods of confinement in orbital space stations or bases on Mars or the Moon, and the problems of extended weightlessness might be countered by the use of some form of artificial gravity.

Once NIFTE reached Titan, the crew would deploy foldout wings on the spacecraft, and the ship would literally cruise through the moon's soupy atmosphere, taking in fuel in the form of methane along the way. As Zubrin points out, "Titan, with one-seventh Earth's gravity and four and a half times Earth's atmospheric density, is an aviation paradise." During the Titan encounter, the crew would survey the moonscape to locate possible sites for a future base or colony and retrieve samples of the surface to see if its organic constituents could be exploited by settlers within an enclosed habitat. Then NIFTE would make for another Saturnian satellite or head home.

Ultimately, colonization beyond Mars will depend on a breakthrough propulsion method such as the magsail that could power a large craft carrying hundreds of settlers to a distant site. Yet even with that advance, the possibility remains that none of the solid bodies in the outer Solar System will prove sufficiently hospitable to induce space dwellers to exchange the security and mobility of orbital habitats for a dubious existence on a forbidding surface. The very density of Titan's atmosphere, for example, could make life beneath its perpetual clouds dismal and dangerous. And potential colonization sites beyond Saturn—within the orbits of Uranus, Neptune, or tiny Pluto—may be ruled out for lack of energy from the Sun, which does little to dispel the gloom on the farthest planets even at high noon. The real lure for humans who dream of settling distant worlds may lie beyond those dark barrier islands of the Solar System and across the ocean of interstellar space. One day, drawn perhaps by reports from a robotic probe of a blue planet circling a neighboring star, bold colonists may attempt the ultimate voyage, to sow the seeds of civilization in the warmth of a new sun.

MINING THE NEW FRONTIER

Among the preeminent challenges facing twenty-first-century spacefarers will be finding cheap, abundant resources to power starships and to build space stations and other structures. Constructing even a modest space dwelling will require millions of tons of materials, yet wresting just a fraction of this from Earth and overcoming the planet's powerful gravitational field to boost the supply into space will cost billions of dollars and squander phenomenal quantities of fuel.

One reasonable alternative is to tap the reserves of space itself: the asteroids. These primordial bodies fall into many classifications, but scientists consider two types—"S" (silicaceous) and "C" (carbonaceous)—to be the most promising candidates for the mining of water, minerals, and carbon compounds. An S-type asteroid measuring half a mile in diameter could contain seven billion tons of iron and a billion tons of nickel. A similarly sized C-type may hold 10 to 20 percent of its weight in water, bound up with minerals such as magnesium, silicon, and sodium.

Asteroids by the millions roam the region between Mars and Jupiter known as the main asteroid belt, too far away to be of practical use for early missions. But a small number—perhaps as many as 1,000— follow odd orbits that periodically bring them near Earth's path. Nearly 140 of these Earth approachers have been identified, and at least five of them can be reached on more fuel-efficient trajectories than those for going to the Moon. Moreover, as illustrated in the scenarios on the following pages, carrying away asteroidal resources to construction sites in Earth orbit will require only a fraction of the energy used to launch from Earth because these tiny rocky bodies have virtually no gravity.

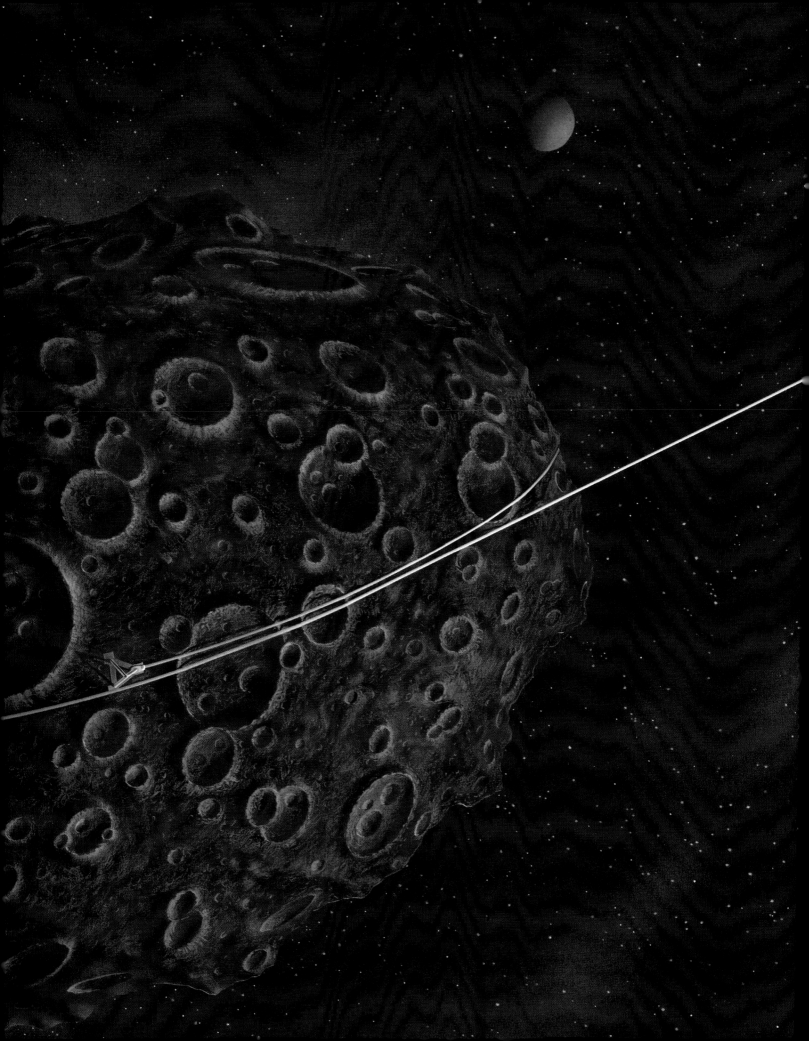

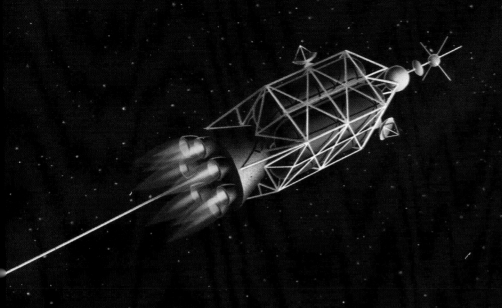

Steadying the Target

Most Earth-approaching asteroids—collectively known as Apollos—rotate once every four and a half hours, a rate that would make excavation difficult because equipment and loosened rock would fly off the surface. One method for halting or slowing this rotation, a process referred to as despinning, pits the asteroid's own motion against itself. As depicted here, a computer-controlled space tug, launched from a mother ship, anchors a lightweight cable to the as-teroid's midsection and then accelerates away, in a direction opposite the asteroid's spin, until the line is slightly taut. As the asteroid continues to rotate, the cable wraps around the equator, like the string on a child's top. When the asteroid has fully lassoed itself, the tug fires its engine and thrusts against the spin until, over the period of about a week, the asteroid ceases its tumbling.

In some instances, despinning will be used only to slow the asteroid's roll, not to stop it. A slight spin can actually assist in certain mining operations, such as the one shown on the next page.

Bagging an Asteroid

The first target of asteroid mining missions will be readily obtainable substances such as water, which could be extracted and then transported to facilities in low Earth orbit, where it would be used for life support or processed into liquid hydrogen-oxygen propellant to boost spacecraft into higher orbits or to fuel long-distance missions. Because the clay minerals of certain C-type asteroids are thought to contain up to 20 percent water by weight, those bodies present the best prospects for this initial phase of mining in space.

In the scenario shown here, a spaceship partially despins a C-type asteroid and then secures a huge impermeable bag over the mining area by a series of cables. The asteroid's slight rotation hurls debris loosened by a surface dredger into the bag, which prevents the mined material from escaping into space. When the bag is full, the spaceship tows it to the focus of a parabolic solar mirror *(above, right)*. There, sunlight heats the bag and its contents to between 1,000 and 1,500 degrees Fahrenheit—hot enough to separate water from its mineral bonds. The resulting water vapor is piped to the ship, where it is condensed and stored in tanks for later conveyance to low Earth orbit.

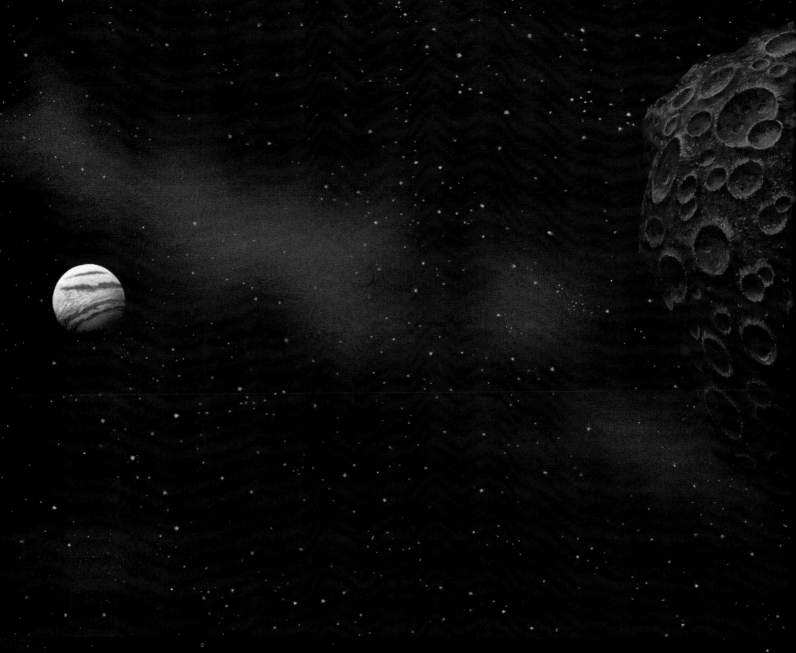

Bringing Home a Rock

Today, lofting ten pounds of iron or nickel into geo-synchronous orbit, 22,300 miles above Earth, costs about $150,000. Carrying the same weight from a mine on a near Earth asteroid to a manufacturing facility in Earth orbit would cost a fraction of that sum—assuming the asteroid mining infrastructure was already in place. Greater savings yet could result if an entire asteroid were brought into Earth orbit, reducing the mine-to-market time from years to days and allowing mining crews to return earthside more often.

Space engineers envision using a type of electric catapult called a mass driver to ferry whole asteroids homeward. Unlike traditional propulsion systems, the mass driver uses no rocket fuel; the Sun, gravity, and the asteroid itself provide the necessary energy. As illustrated above, the telerobotically operated mass driver *(tubelike structure)* attaches itself to the aster-oid—in this case an iron-rich S-type—and digs out a quantity of fine rock. Internal electric coils, powered by solar panels, generate a magnetic field that accel-erates the asteroid material and propels it at enor-mous speed down the driver's half-mile-long conduit and into space. The force of expulsion thrusts the billion-ton asteroid on a Venus-Earth trajectory. Once it reaches cruising speed, the mass driver shuts down and the asteroid coasts, making course changes with a gravity-assist from Venus *(above, left)*. Upon arrival, the driver parks the asteroid in geosynchronous orbit where it can be easily mined.

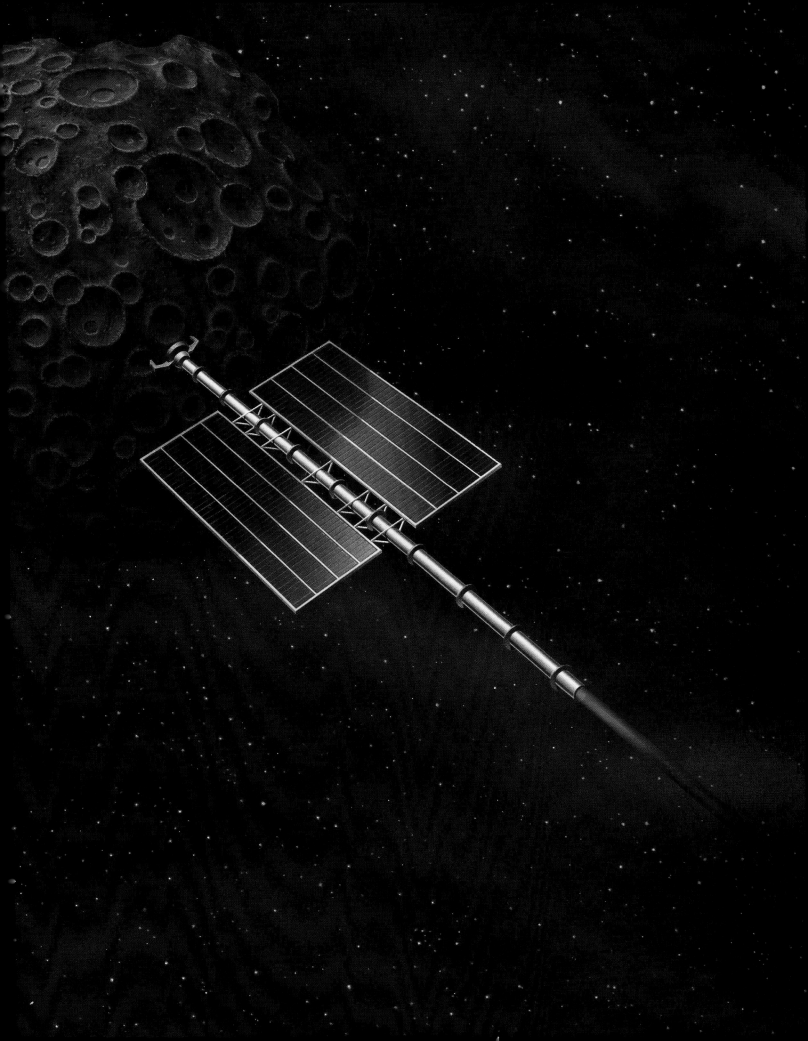

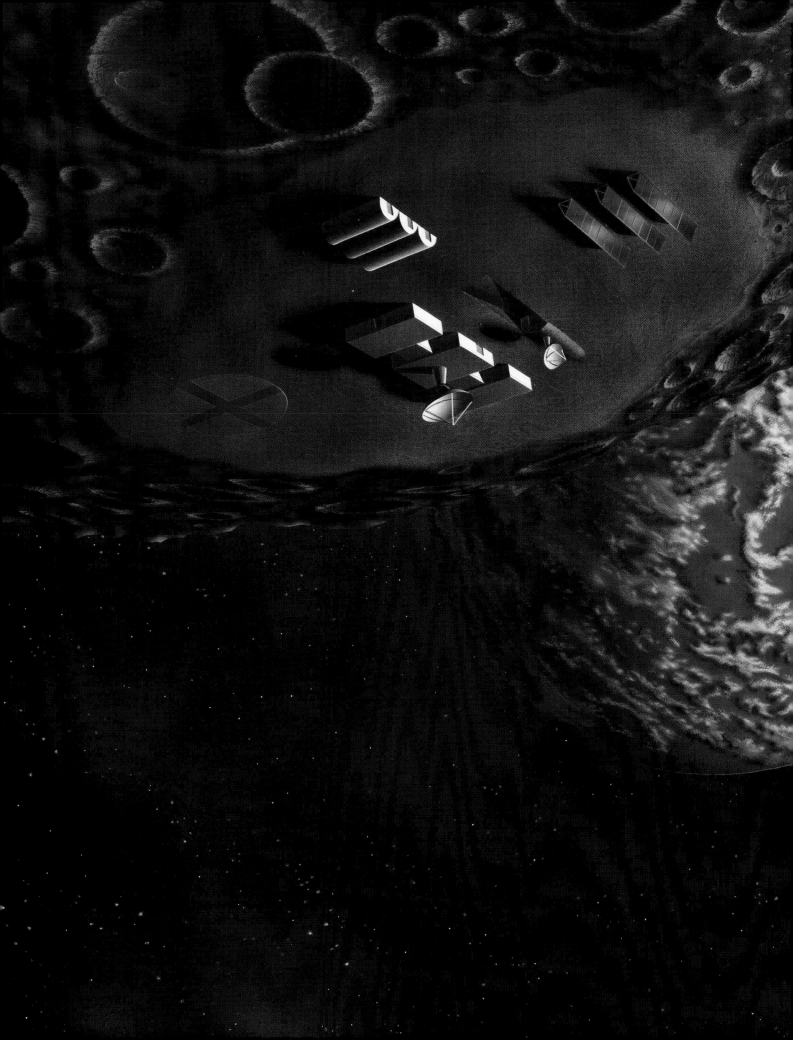

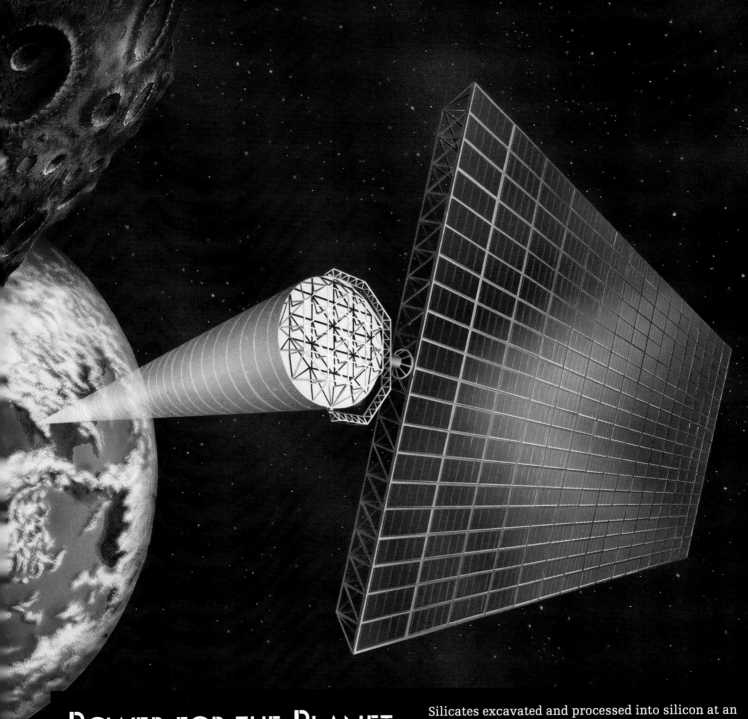

Power for the Planet

A half-mile cubic chunk of metal-bearing asteroid delivered to Earth orbit would be a boon to space industrialists and power utilities alike. At today's consumption rates, just one such asteroid could provide a fifteen-year supply of steel and enough nickel to meet planetary requirements for two hundred years, and its bounty of resources could be used to manufacture starships, lunar modules, and space habitats. Or, by converting only a small portion of the asteroid's raw materials into a solar-power satellite *(above, right)*, space engineers could provide Earth with a clean, economically competitive source of electricity.

Silicates excavated and processed into silicon at an S-type asteroid mining facility *(upper left)* could be used to fashion the photovoltaic cells that make up the satellite's six-mile-long solar panel. Supporting struts and beams and the satellite's half-mile-diameter transmitting antenna could be produced from the asteroid's supply of iron and nickel.

Once constructed and deployed, a satellite such as this one would generate five times the power of a nuclear plant. Sunlight striking the satellite's solar cells would be converted to electricity, which would in turn be transformed into a focused beam of microwaves and transmitted to Earth. Received and reconverted to electricity, the energy could supply some three million homes a year.

The young stars of the Pleiades cluster, shown here in an optical photograph, inspire hope among scientists who are searching for habitable worlds beyond the Solar System. Infrared astronomers have detected what may be protoplanetary disks or spheres around a few of the stars, leading some to suspect that older suns may host more developed systems—complete with planets hospitable to life.

ecade upon decade, the small robotic probe had cruised silently through the vast interstellar darkness, pressed ever onward by the feathery but incessant touch of a laser beam against the silver of its sail. Then, like a model airplane on the end of its wire, it ceased its outward flight—captured by the gravity of a mid-size orange star known to terrestrial astronomers as Epsilon Eridani. Now, the probe's looping trajectory has dropped it into orbit around a medium-size planet circling its sun at a distance of just under 50 million miles. While the orbiter takes a series of initial readings of the planet's atmosphere and gravity, it dispatches a team of mobile robot landers to various spots on the surface, some 300 miles below. With insectlike diligence and agility, the landers roam the surface, taking soil samples for chemical and mineral analysis, looking for water, and scouting for signs of life.

Sixty-three and a half trillion miles away on Earth, more than a hundred scientists, engineers, and project managers—many of whom were not yet born when the probe was launched—stand watch in mission control, reviewing the latest data and struggling to contain their impatience. Every message from the orbiter at Epsilon Eridani is more than ten years old by the time it arrives, and the knowledge that anything might have happened since it was sent only sharpens the mix of anxiety and excitement permeating the room.

Certainly the day has been long in coming: Generations of astronautical dreamers have been building toward it since the late twentieth century, when observers began discovering tantalizing hints that the Sun was not the only star in the galaxy that might have hatched a clutch of planets. Now, if the robot emissary reports that this promising Eridanian planet is habitable, the most incredible migration in the long history of humankind will begin. . . .

By the time this scenario can come about, many of Earth's children might have abandoned the security of their home planet for life in orbiting artificial colonies or in domed shelters on the Moon or Mars. But before anyone dares to brave the black void beyond the Sun's domain, they must have a destination—or at least a reasonable hope of one. For now, hope is a glimmer.

Data gathered in 1983 by a relatively short-lived orbiting telescope known as the Infrared Astronomical Satellite *(IRAS)* sparked tremendous excitement among those searching for extrasolar planets. Analysis of the *IRAS* results indicated that a smattering of stars in different parts of the sky produce more infrared radiation than expected for stars of their type, age, and size. Long-wavelength infrared, which is emitted by anything with a temperature above absolute zero, is perhaps the most common form of energy in the cosmos. But because stars are very hot, they usually radiate most strongly at much shorter wavelengths. Astronomers therefore speculate that the excess infrared picked up by *IRAS* is being given off by a disk or shell of dust particles around each of the stars in question. Scientists believe that such a circumstellar disk of material, remnants of the gaseous nebula out of which the star formed, represents an evolutionary stage in the life cycle of many stars and is a precursor to the formation of planets.

This is good news for would-be starfarers: Planetary systems akin to the one that formed around the Sun some 4.6 billion years ago may well be the rule rather than the exception in the universe. But finding such a system within striking distance of Earth is easier said than done. The Sun's nearest stellar neighbor, at 4.3 light-years away, is actually a trio of stars, the Alpha Centauri system, orbiting a common center of mass. Multistar systems occur frequently in the Milky Way (there is even serious speculation that the Sun may have a very dim and distant companion), but the gravitational interactions among the members of such groupings could in some cases preclude the development of protoplanetary disks and, thus, of planets. Some scientists suggest, however, that in this case the three stars—Proxima Centauri, Alpha Centauri A, and Alpha Centauri B—are in fact far enough apart so as not to interfere with planetary formation. Furthermore, other researchers, citing Alpha Centauri A's resemblance to the Sun in terms of mass and temperature, think it quite probable that that star at least could possess a planetary system. Infrared astronomers cannot say for sure, though, because the stellar triplets lie on a line of sight from Earth that puts them directly in front of the heart of the Milky Way; the glare from the galaxy's star-packed center makes it difficult to read the trio's infrared signatures.

The next nearest potential destinations are much farther away than Alpha Centauri. Since the mid-1980s, Dana Backman of NASA's Ames Research Center at Moffett Field, California, has been studying 134 Sun-type stars within a radius of eighty light-years of Earth. At least 25 of the stars emit the telltale excess infrared radiation that may be from circumstellar material, and Backman has suggested that the actual number may be much higher. "If *IRAS* had been more sensitive," he has argued, "we could probably detect remnants of planet-forming debris—although not actually planets themselves—around every one of those stars." Backman's research indicates that the protoplanetary debris does not extend all the way in to the star but tends to lie at a distance equivalent to the distance of Pluto from the Sun, or farther, implying that debris starward of that orbit might have been swept up by planets as they formed.

Epsilon Eridani, which lies 10.8 light-years away in the constellation Eridanus, is an especially intriguing star in Backman's survey. Not only does it show the excess infrared signature of surrounding debris but it also seems

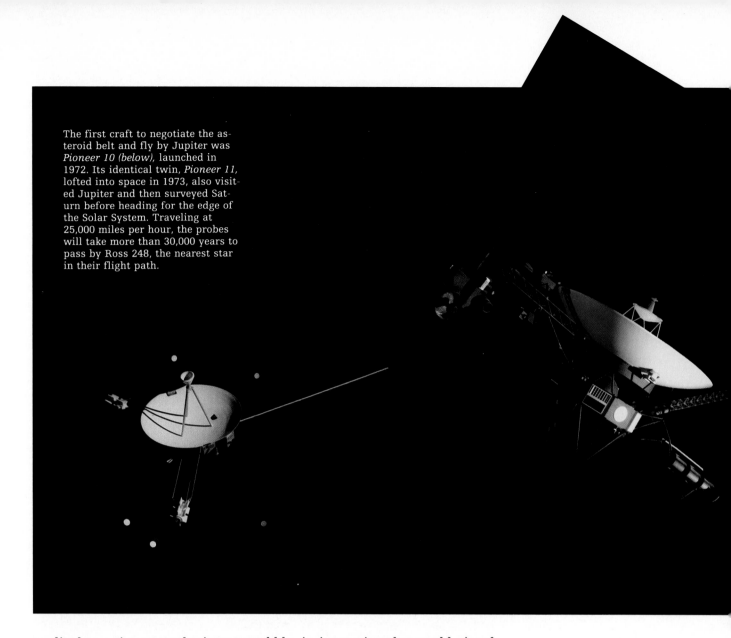

The first craft to negotiate the asteroid belt and fly by Jupiter was *Pioneer 10 (below),* launched in 1972. Its identical twin, *Pioneer 11,* lofted into space in 1973, also visited Jupiter and then surveyed Saturn before heading for the edge of the Solar System. Traveling at 25,000 miles per hour, the probes will take more than 30,000 years to pass by Ross 248, the nearest star in their flight path.

to display a tiny perturbation, or wobble, in its motion that could signal gravitational interaction with another orbiting body or bodies. The wobble, a preliminary finding by Canadian astronomer Bruce Campbell and his colleagues, could be caused by instabilities in the star itself, however, and until Campbell's measurements are complete, the question will remain unresolved. But if planets have already formed around Epsilon Eridani, it is possible that one of them might have come into being in the so-called zone of habitability, just far enough from the star to have developed a moderate range of temperatures. If such a planet was also a terrestrial body, like Mars and Earth, rather than a gaseous type like Jupiter and Saturn, then the chances increase that the planet is suitable for human occupation.

ACROSS THE GREAT DIVIDE

Assuming that scientists like Backman, Campbell, and their successors over the next several decades can amass enough evidence to justify closer scrutiny of Alpha Centauri A or Epsilon Eridani, the next step will be to send robotic scouts. Designing machines that are capable of crossing the inconceivable

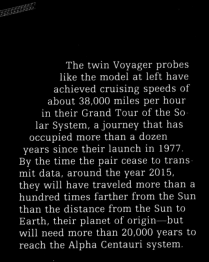

The twin Voyager probes like the model at left have achieved cruising speeds of about 38,000 miles per hour in their Grand Tour of the Solar System, a journey that has occupied more than a dozen years since their launch in 1977. By the time the pair cease to transmit data, around the year 2015, they will have traveled more than a hundred times farther from the Sun than the distance from the Sun to Earth, their planet of origin—but will need more than 20,000 years to reach the Alpha Centauri system.

distances involved is the first—and perhaps the most difficult—challenge ahead. Light itself, traveling at 186,283 miles per second, takes 10.8 years to reach Earth from Epsilon Eridani, and no ship yet built can approach even a significant fraction of the speed of light.

The most far-ranging craft yet to leave from the Sun's third planet are four unpiloted probes that were catapulted into space during the 1970s to gather information about the giant bodies of the outer Solar System. *Pioneer*s *10* and *11* and *Voyager*s *1* and *2*, having completed their primary missions, are now heading in different directions toward the interstellar frontier. The twin Voyagers are cruising at about 38,000 miles per hour, the slightly older Pioneers at 25,000 miles per hour, both phenomenal speeds by terrestrial standards. But none of the four will come within hailing distance of the nearest stars in their flight paths for tens of thousands of years.

Fast as they are, and as sturdy as they have proved themselves to be, the Pioneers and Voyagers are not destined to be humanity's ambassadors to the stars. Any such craft will need propulsion systems other than the conventional chemical rockets used to launch today's spacecraft. The central problem with chemical fuels, usually liquid hydrogen and liquid oxygen, is their relative inefficiency in terms of pounds of thrust per pound of fuel. To accelerate to the speeds necessary to escape from the Solar System and make an interstellar crossing in something less than a hundred years, a spacecraft would have to continue to accelerate even after escaping the Sun's gravitational field. Continuous acceleration in turn would require the ship to carry so much fuel that the system would rapidly reach a point of diminishing returns: The more fuel the ship carried, the greater the mass the engine would be trying to propel; in effect, the fuel would be burned to propel more fuel. Moreover, this is only half the problem. In order for the craft to do something more than simply take a quick series of photographs in passing, it will have to decelerate upon nearing its target—a process that requires yet more fuel.

Over the last half-century, a number of visionaries have proposed alternatives to chemical fuel. California experimental physicist Robert Forward, for example, champions a propulsion scheme that is a variation on the solar sail, operating on pressure exerted by a highly focused beam of light generated by a laser. Because sunlight radiates in all directions, the number of photons striking a square centimeter of sail decreases proportionally with distance according to the inverse-square law—that is, when the sail's distance from the source has doubled, the pressure of sunlight, already whisper-soft, is four times less. At the outer edge of the Solar System, the pressure of photons on a sail is reduced by a factor of 1,500 and drops by trillions even halfway to the Alpha Centauri system. Most scientists thus regard solar sails as practical only for interplanetary travel within the orbit of Jupiter.

A laser, in contrast, can concentrate photons in a narrow beam that can travel millions of miles without significant spreading, maintaining its ability to exert pressure—theoretically at least—over the enormous reaches between nearby stars. Of course, no ordinary laser on Earth has anything approaching the power to perform work on such a scale. Forward's design thus calls for an array of thousands of lasers, probably solar-powered, placed in orbit around the Sun near Mercury. The beams from these devices would be combined and focused by a lightweight plastic lens 600 miles in diameter floating in space between Saturn and Uranus. According to some calculations, an early, less powerful version of the laser could be used to accelerate a robot probe weighing 863 tons to an average velocity 10 percent of the speed of light, a rate that would bring it to Epsilon Eridani in about a hundred years.

Another variation on the idea is the microwave sail. A solar-powered satellite orbiting Earth would convert sunlight to electricity, which in turn is used to generate microwaves. In an arrangement similar to that for the laser sail, the microwaves are focused through a lens and beamed at a sail made of aluminum wire mesh. The mesh structure reduces the sail's mass while still providing a reflective surface, and the metallic wire can act as an antenna for communication. Miniaturized infrared, ultraviolet, and other detectors would stud the sail, allowing it to transmit high-resolution images during its journey. A very small probe weighing less than an ounce with a microwave sail .4 miles square could reach Epsilon Eridani in roughly 53 years. For now, microwave technology may have an edge over optical laser technology in terms of being more nearly within reach for space uses, but scientists regard both options as worthy of continued examination and development.

THE NEED FOR SMART MACHINES
At interstellar distances, meaningful control from Earth of robot probes becomes impossible. Even at the speed of light, radio messages to and from a craft investigating the Alpha Centauri system would take more than eight years, clearly an impractical time-lag for responding to emergencies or other unexpected events. One solution would be to build a probe sophisticated enough to deal with the unexpected.

Researchers in artificial intelligence, as this branch of computer science is known, have been pondering the difficulty of teaching computers to emulate human thinking for more than forty years, and although no one would claim to have accomplished this feat, modern computers can in fact do considerably more than simply crunch numbers. Programs called expert systems, for example, can mimic the decision-making skills of specialists in different fields.

To create the systems, programmers interview human professionals to build a database of essential information and distill the experts' logical thought processes when solving problems; all of this is then translated into computer code. Medical expert systems, for example, can diagnose disease by asking patients a series of increasingly focused questions based on answers to more general preceding ones. Another program, developed for the United States Navy, coordinates the movements of 300 ships and 2,000 planes in an area covering more than 95 million square miles of ocean and 2,450 ports. NASA already has expert systems designed to diagnose malfunctions on spacefaring vehicles; assuming adequate funding, such a system could be ready for use on the space station *Freedom* by the mid-1990s.

For purposes of exploring the worlds in orbit around an alien star, however, the on-board system would have to be a kind of Renaissance robot, equipped with expertise from such disparate fields as planetary geology, atmospheric physics, exobiology, and astrophysics. Ideally, moreover, the probe would also be capable of learning from its experiences—functioning, in other words, much like a human brain.

Such a computer would need to operate in something other than the step-by-step fashion of conventional serial machines, which perform even the most complex tasks by processing one piece of data at a time. The approach works well enough for certain undertakings—indeed, serial processors generally far surpass the human brain when it comes to speed and accuracy in calculating. But the machines are not very efficient at recognizing patterns and making comparisons, processes that are critical to interacting with the external world and to learning. Computer scientists are thus experimenting with so-called neural networks, a circuitry design based on the physical organization of the human brain. To fully duplicate the brain's complexity is beyond reach for now; it has billions of neurons and perhaps a thousand times as many connections between neurons. However, smaller-scale neural networks resemble the brain in one important respect: They operate in parallel. That is, instead of one processor handling data in sequence, many processors, or "neurons," tackle information arriving simultaneously along multiple pathways. Furthermore, because each processor is linked to all others, the machine can solve problems by a continuous but very rapid trial-and-error process, the result of innumerable simultaneous calculations. In human terms, the process is similar to the way the brain takes in countless pieces of sensory information and combines them with equally countless pieces of stored information—conscious and subconscious—in order to arrive at a decision.

A different parallel-processing approach to building adaptable robots has been taken by researchers at MIT's artificial intelligence laboratory. Instead of trying to replicate the human brain with miles of connecting circuitry and hundreds or thousands of processors, a team headed by Rodney Brooks has devised small, lightweight robots programmed to follow simple commands that nevertheless allow the machines to perform a variety of useful jobs.

The key to the robots' versatility lies in their hierarchical programming,

A Trio for Robotic Reconnaissance

The first reconnaissance missions to planets around a distant sun are liable to be carried out not by men and women but by robots. Such scouts will not need huge amounts of oxygen, food, and water for the round-trip journey, which in any case may last many times longer than a mortal lifetime. To pave the way for later human explorers, these early investigators will have to do a wide range of useful work on the surface. Unlike the two Viking probes, which could take soil samples only within reach of where they touched down on Mars, devices similar to the three prototypes described here—designed and built by researchers at MIT's artificial intelligence laboratory—could be pro-

grammed to range far afield to gather information.

The premise underlying the creatures' insectlike design is that their programming can be arranged in hierarchical layers that combine to produce complex behaviors. For example, Genghis, the six-legged robot below, is programmed to carry out a series of movements, beginning with simply standing upright, that result in walking. Its next level of competence allows it to traverse rough terrain, and its next directs it to pursue heat sources, such as people. As different behaviors interact with one another in response to various stimuli, Genghis and its cousins can react quickly to their surroundings.

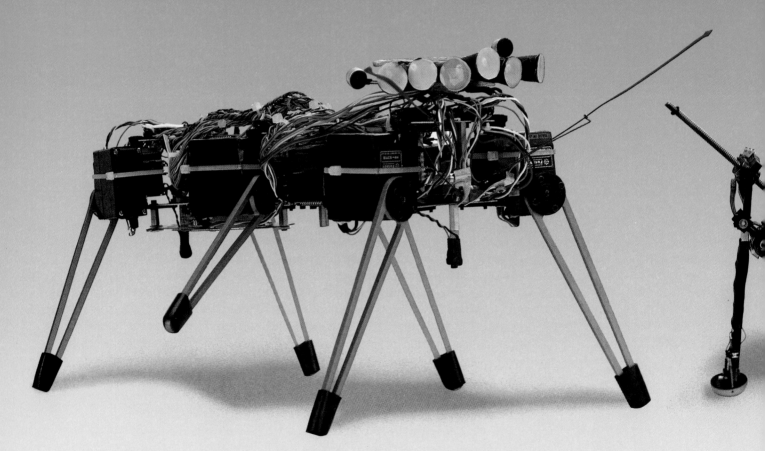

A 1.3-cubic-inch wheeled device known as Squirt, shown life-size at left, is a kind of eavesdropping bug, programmed to scurry away from light and toward sound. Though not specifically designed for space missions, Squirt demonstrates the potential for robot miniaturization. For example, an army of such minicreatures could easily be dispatched to map the surface of another planet.

Genghis, a foot long and six inches high *(above)*, is a model of decentralized programming. Of the creature's nearly sixty behavior programs, only four coordinate the overall functioning of its legs as a group; eight programs operate independently on each leg. Another pair of programs cause the robot to track people with eyelike infrared sensors.

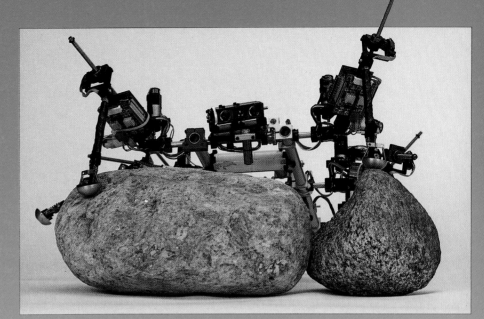

Attila *(below)* is a robot about the same size as Genghis but designed with many improvements. For instance, its jointed legs give it more flexibility for scaling obstacles such as rocks *(right)*. Moreover, it can measure the inclination of a slope within two degrees—useful for mapping—and has an inertial guidance system that enables it to return to its starting point. As a prototype planetary rover, Attila is equipped with minicameras and sensors in its footpads to detect obstacles in its path as well as the texture and firmness of foreign soil.

which results in so-called emergent behavior. For example, the robots are programmed to perform a series of movements: "lift leg," "swing leg forward," "balance forces on all legs" (which, if one leg is in the air, causes the robot to put that leg down). Although the movements are not designed to occur in any particular order, the "lift leg" behavior overrides all the others; thus, if all legs are on the ground—and all forces are balanced—"lift leg" is triggered, to get the robot moving again. In effect, then, the complex action of walking "emerges" from the combination of simple directives. The result is a handful of mobile, insectlike devices that could be used, for example, to gather data on the surface of an alien planet *(pages 100-101).*

EXODUS

Once reconnaissance of an extrasolar planet is complete and would-be colonists have learned all they can about their future home, their extraordinary journey can begin. According to astronautical engineers, the vessels that will carry the seeds of humanity starward could be propelled by any one of several systems capable of accelerating a ship to velocities ranging from 10 percent to 50 percent of the speed of light. At the upper end of this range, spacefarers could expect to complete the journey—barely—within their lifetimes. At the lower end, provisions would most likely have to be made for several generations of pioneers to coexist aboard ship, with all that would entail, including rearing and educating children born in space.

Many of the propulsive methods have been on the drawing boards for decades. Some expand on proven technologies, and others have so far lived only in theory. For example, although liquid hydrogen and liquid oxygen are out of the running as fuel for interstellar flight, another form of hydrogen may give the chemical rocket new life. In 1989, researchers at the Carnegie Institution of Washington reported an intriguing breakthrough when they placed gaseous hydrogen in a diamond anvil and subjected it to millions of times normal atmospheric pressure. The preliminary result appears to be metallic hydrogen, a solid, incredibly dense form of matter that does not exist on Earth (although it may be a significant constituent in the makeup of Jupiter).

Provided that metallic hydrogen is stable at normal pressures, it could theoretically pack so much potential energy into such a small mass that it could be used to accelerate a starship carrying a human crew and the necessary life-support systems (a payload estimated at 500,000 pounds) to 20 percent of light-speed over a period of a few years, with a dramatic reduction in the amount of fuel needed for interstellar flight. Such a ship could reach Epsilon Eridani in fifty-nine years, a journey that includes three years spent accelerating to the ship's cruising speed and an additional three years decelerating to go into orbit around its target.

The seemingly leisurely years of acceleration and deceleration are of vital importance to a ship's crew. Human beings would be severely uncomfortable after more than a few days of exposure to gravity greater than 1g, the force exerted at Earth's surface. (At launch and reentry, shuttle astronauts are presently subjected to forces ranging from 2g to 6g.) And such exposure would also have less fleeting detrimental effects. Even a few seconds at 20g, for instance—the pressure astronauts have tested in a special centrifuge—causes capillaries to burst; the forces engendered by rapid acceleration to, or deceleration from, any large fraction of light-speed would be fatal.

Another propulsion scheme involving unusual forms of matter is known as the matter-antimatter engine. According to nuclear physics, each subatomic particle of matter has an antimatter counterpart identical in mass but opposite in such properties as charge and direction of spin. The antiproton, for example, is the negatively charged counterpart of the positively charged proton, and the positron is the positively charged antiparticle to the negatively charged electron. When such pairs collide, they annihilate each other, producing a tremendous burst of highly energetic gamma ray photons.

In the 1950s, intrigued by the idea of a fuel that could achieve an exhaust velocity equal to the speed of light, German rocket scientist Eugen Sänger proposed the idea of using matter-antimatter annihilation to propel a spaceship. Although Sänger was a leading pioneer in the field of rocketry and space travel, his proposal sparked only mild interest among his peers, largely for lack of a way to turn theory into practice. The problem was that the gamma rays produced in collisions between electrons and positrons (the only form of antimatter that had been discovered at the time) dispersed in mere fractions of a second and in all directions; no one could figure out how to harness their energy. In subsequent experiments with powerful particle accelerators, however, scientists have found that collisions between protons and antiprotons produce what could be a useful intermediary, a short-lived particle called the pion, which appears briefly and then rapidly decays into gamma rays.

Robert Forward among others has suggested techniques for harnessing the energy of matter-antimatter collisions by capturing pions before they decay. This is possible because some of the pions, unlike gamma rays, are charged particles and can thus be trapped like fireflies in a kind of electromagnetic bottle. In theory, a spacecraft supplied with antimatter in a magnetic field antimatter storage facility could force matter and antimatter particles together, using a specially designed series of additional magnetic fields to generate pions and focus them into a high-energy exhaust stream. Such a system could propel the crew-carrying ship to 20 percent of light-speed, bringing the Eridanian pioneers to their new world in only fifty-seven years (two years fewer than the journey in a ship propelled by metallic hydrogen because less time is needed for acceleration and deceleration).

For the moment, the projected cost of manufacturing antimatter—$100 billion per milligram—makes such a propulsion system prohibitively expensive. But the energy yield is so enormous that Robert Forward, at least, has argued that matter-antimatter engines could be competitive with chemical rocketry as soon as the cost drops below $10 million per milligram. By the time other practical barriers to interstellar flight are removed, the pion rocket may make economic sense.

AND NOW, THE ATOM

Perhaps more achievable in terms of existing technology are interstellar arks powered by nuclear fission, the splitting of the atom that led to the atomic bomb and fuels modern nuclear power plants. In nuclear power plants, atoms of radioactive elements such as uranium or plutonium are split in a chain reaction that liberates heat, which is used to turn water into steam, which in turn drives electric generators. In space, the reaction would heat not water but a gas—probably hydrogen, the most abundant and lightest-weight element in the universe. As its temperature increased, the hydrogen would rapidly expand and spurt out of a rocket nozzle with a thrust capable of propelling a starship to five percent of light-speed or greater.

For greatest efficiency on interstellar flights, the hydrogen should reach maximum temperature quickly, to develop an exhaust stream with the greatest possible thrust. However, with solid radioactive fuels such as those used in terrestrial power plants, the fission reaction must be carefully regulated to avoid overheating the fuel rods to the point of meltdown. Slowing the reaction also reduces the reactor's energy output, however. Astronautical engineers have thus considered a nuclear engine that would use uranium or plutonium fuel that is already liquid, or even gaseous. With fear of meltdown no longer a constraint, the reactor could be allowed to get much hotter, which would heat the propellant more quickly and to a higher temperature. As a propulsion method for starships, a gaseous-fuel fission reactor is not as efficient in generating propulsive thrust as an engine powered by metallic hydrogen or matter-antimatter, but—given technological advancements—it could accelerate the starship to 15 percent of light-speed, a rate that would bring the vessel to Epsilon Eridani in seventy-five years.

Yet another propulsion design based on the technology of the atomic bomb is the nuclear-pulse rocket. Its premise is simple: Every few seconds, a fission bomb is detonated a short distance behind a starship equipped with an enormous metal pusher plate attached to the ship's aft section by equally enormous shock absorbers. Vaporized debris from the explosion slams into the plate, compressing the shock absorbers. With each detonation, the shock absorbers compress and rebound, pushing the ship forward in pulses. Fanciful as it sounds, a miniature prototype, powered by conventional chemical explosives, proved the theoretical soundness of pulse propulsion in 1959 by flying for several hundred feet over the desert in southern California.

The device has a certain pragmatic appeal: It could be built tomorrow with current technology, provided it were constructed and launched in deep space to avoid contaminating Earth's atmosphere with nuclear fallout. But the nuclear-pulse rocket scores very low on the speed scale, accelerating a crew-carrying ship to only three percent of light-speed in four years. The vessel's slowness is a function of several factors, related in a kind of chicken-and-egg conundrum. First, the weight of the bombs and of the necessary radioactive shielding adds considerably to the overall weight of the spaceship (thus requiring more power for acceleration). In addition, repeated bomb blasts will

MESSAGES AMONG THE STARS

Among the challenges facing star voyagers of the future will be keeping in touch with the folks back home. Electromagnetic waves, such as radio waves and microwaves, will likely remain the communications medium of choice. Traveling at the speed of light, or 186,283 miles per second, these signals are second to none in terms of velocity and are easy to generate and to detect. However, although they can pass relatively unimpeded through interstellar clouds of dust and gas, they are subject to some interference and are blocked by solid objects such as planets and stars; some messages would be garbled or might not get through at all.

Future communicators may thus attempt to develop alternatives, one of which might involve gravity waves. Einstein's theory of general relativity describes gravity waves as infinitesimal ripples in the fabric of space-time caused by the violent motion of extremely massive objects such as black holes. They propagate at light-speed and would offer the distinct advantage of traveling completely unobstructed from transmitter to receiver. But in order to generate them, engineers would have to be able to create or harness incredibly dense accumulations of matter and then control their motions to encode signals. Reception would prove equally difficult, given that scientists estimate gravity waves disturb the space-time fabric by as little as a billionth of a trillionth of a centimeter—thousands of times beyond the capabilities of today's most sensitive detectors.

Neutrinos *(above)* represent another option. These subatomic particles are produced by nuclear fusion reactions and radioactive decay and would be relatively simple to generate. They travel at light-speed and consist of little or no mass, so they can pass unscathed through solid objects. This characteristic, however, makes them notoriously difficult to detect.

Even if such problems can be overcome, none of these methods will be practical for conducting conversations with far-off spacefarers or, perhaps, alien cultures. Earth's closest stellar neighbor, Proxima Centauri, is 4.3 light-years away; one single transmission and reply would take nearly nine years to complete. As explained on the following pages, trading information over even greater distances may require some fundamental changes in communications strategies.

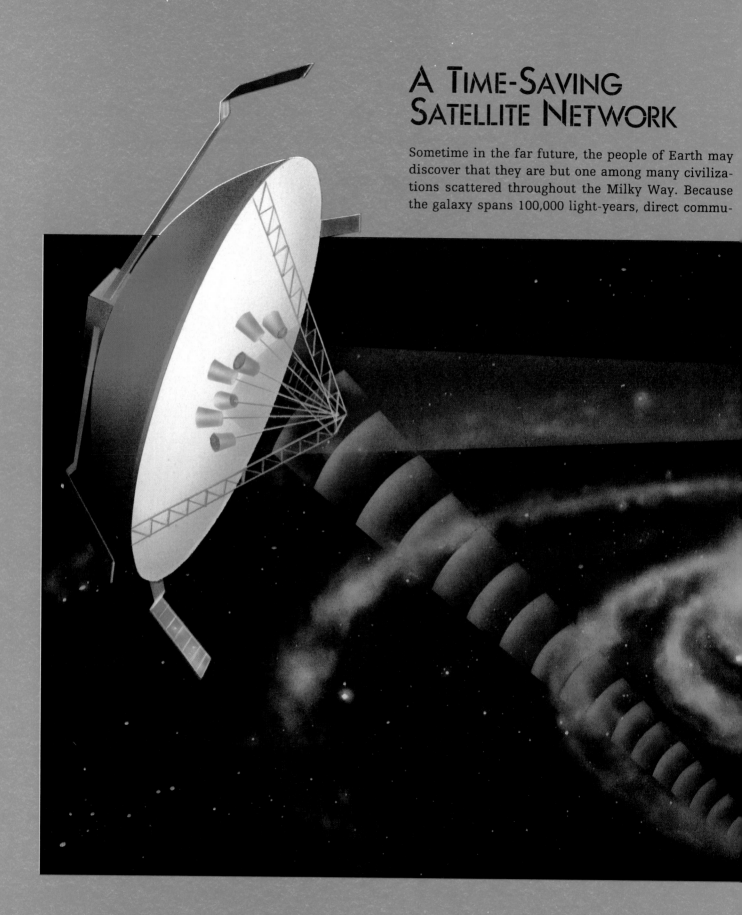

A Time-Saving Satellite Network

Sometime in the far future, the people of Earth may discover that they are but one among many civilizations scattered throughout the Milky Way. Because the galaxy spans 100,000 light-years, direct communications will be agonizingly slow. Following initial contact, however, far-flung societies might agree to establish a communications network that could significantly speed things up. Each world would be responsible for building a satellite station *(below)* equipped with powerful antennas that would transmit and receive signals—most probably at microwave frequencies—and a computer database with exceptional storage capacity. Locations for the satellites would be carefully selected to minimize transmission times from station to station and from each station to its home society. A civilization would then use its local satellite database as a repository for all information about itself, and the various satellites would transmit this knowledge to one another *(solid signal)*. With the accumulated wisdom of the galaxy shared throughout the network, queries about distant civilizations could simply be directed to the nearest communications station *(pulsed signals)*.

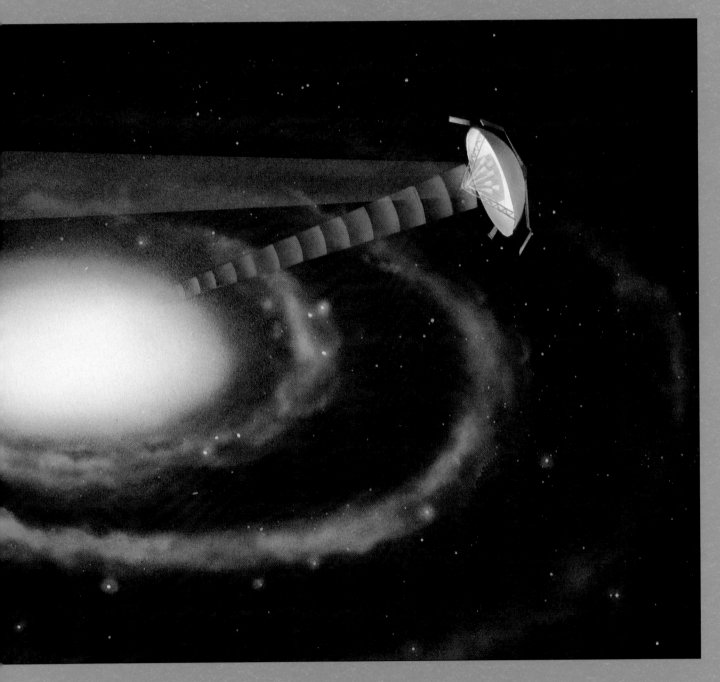

Beyond the Ultimate Barrier

Although perhaps an unachievable dream, faster-than-light communication is at least theoretically possible. According to special relativity, massless phenomena such as electromagnetic waves can travel at light-speed, but ordinary matter can never reach such a velocity because its mass approaches infinity as its speed increases. However, the mathematical equations of relativity allow for the existence of an-

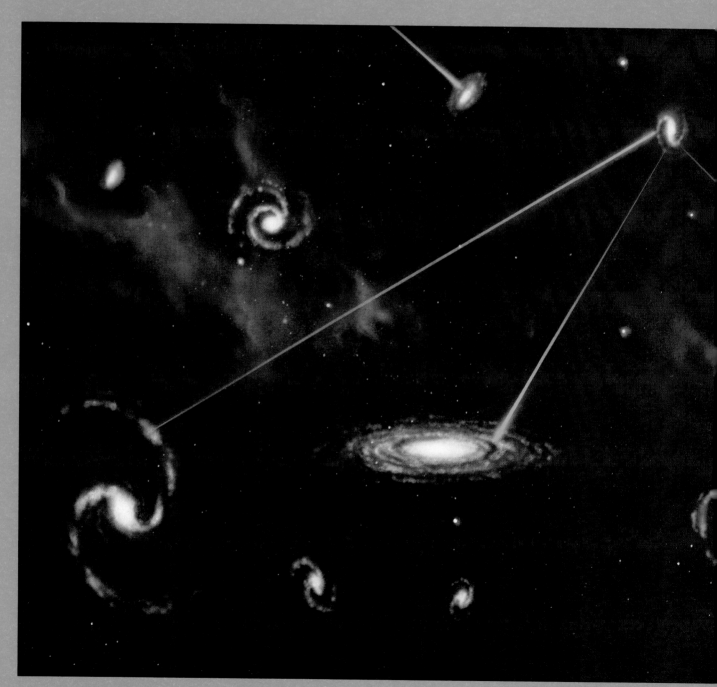

other form of matter on the opposite side of the light-speed barrier. These hypothetical particles, known as tachyons (from the Greek word for swift), possess so-called imaginary mass. Tachyons could therefore never travel at or slower than light-speed but would have no upper limit on their velocity.

Physicists still do not know if tachyons actually exist; experiments designed to detect these particles have been unsuccessful and ideas for generating them are only at the speculative stage. But if these super-luminal particles turn out to be real and if they could somehow be harnessed, virtually instantaneous communications would be possible—not only within the Milky Way but all across the universe, among galaxies that are separated by millions and even billions of light-years *(below)*.

ultimately degrade the pusher plate and will also require periodic safety checks for radiation seepage; while these checks occur, the explosions will have to stop, interruptions that add to the overall length of the journey. Then, because a journey to Epsilon Eridani at three percent of light-speed will take about 330 years, the ship must be prepared to accommodate some ten generations of travelers—which again increases the payload.

FUSION INSTEAD OF FISSION

Considerably more impressive—and perhaps more appropriate for interstellar travel—is an engine fueled by the fusion of atomic nuclei, reactions similar to those that power the Sun and all its cousins. In stellar fusion, energy is released when hydrogen nuclei are fused by the phenomenally high pressures at the core of stars. In a fusion reactor such as one that might power a starship, a plasma containing isotopes of lightweight hydrogen, helium, or lithium would have to be heated to at least 100 million degrees Kelvin.

The process has several stumbling blocks. First, scientists have not yet succeeded in initiating, sustaining, and manipulating the energy from a nuclear fusion reaction. The hydrogen bomb—so far the only device built that employs nuclear fusion—is not an efficient model for a propulsion system: Less than 20 percent of the energy released would be usable; the rest escapes in the form of lethal short-wavelength radiation. Simply to get the plasma hot enough to start the fusion reaction presently requires an expenditure of more energy (to generate the necessary magnetic fields and laser beams) than the reaction itself produces.

Discouraging as these obstacles may seem, some futurists have not given up on a fusion engine. In 1990, scientists at Princeton University's plasma physics laboratory reported having nearly achieved the break-even point. In theory, if a means could be found to contain the process safely and reduce the energy costs of start-up, a ship propelled by a fusion system with a high ratio of energy per unit of mass could reach a whopping 30 percent of light-speed, cutting the voyage to Epsilon Eridani to thirty-nine years.

Whatever the source of their energy—be it metallic hydrogen, matter-antimatter, fission, or fusion—all of these propulsion methods have not advanced beyond the conventional chemical rocket in one significant respect. A spacecraft powered by any of them would need to have its entire fuel supply for the outbound journey, and a similar amount for any planned return, on board when it was launched from Earth, a necessity that imposes an upper limit on the craft's speed. That is, adding fuel tanks in order to be able to go faster ultimately becomes counterproductive since it increases the overall mass that needs to be accelerated.

As long ago as 1960, one imaginative solution to the problem occurred to Robert Bussard, then a physicist at Los Alamos Scientific Laboratory (now called the Los Alamos National Laboratory). Bussard named his device the interstellar ramjet; he might well have called it an interstellar vacuum cleaner. Although space is a vacuum more nearly perfect than can be created in terrestrial labs, it is not altogether empty. Every cubic inch of the void contains a few atoms of hydrogen and occasional ions, nuclei that have been stripped of some or all of their electrons. Although scant, in the trillions of miles between stars they would add up to a bountiful source of fuel for a fusion rocket. To scoop up this free-floating fuel, Bussard proposed to generate a cone-shaped magnetic field just under 2,000 miles across at its widest. Pointing in the direction of travel, the magnetic scoop would collect the hydrogen in its path and funnel it into a nuclear fusion reactor engine.

With no need to carry excess fuel and what Bussard estimated to be a nearly unlimited supply en route, the physicist speculated that the ramjet could accelerate to light-speed itself. More recent calculations, based in part on considerably lower estimates of the abundance of hydrogen atoms between stars, have capped the ramjet's potential at 50 percent of light-speed—a more than respectable velocity. At this rate, the travelers to Epsilon Eridani would reach their destination after twenty-eight years aboard ship.

INTERSTELLAR ARKS

Given the ramjet's capacity for sustained acceleration (and deceleration), most of the crew aboard a ramjet-equipped spacecraft could reasonably expect it to reach the Eridanian planet within their lifetimes. Metallic hydrogen or matter-antimatter engines would lengthen the trip by some thirty years, a very long voyage, but—conceivably—still within crew members' life spans if they were in their late teens, say, at the outset. However, the engineering breakthroughs needed to build propulsion methods capable of these speeds lie far in the future, and the human species is not noted for its patience. Rather than wait for superfast ships, would-be stellar travelers may opt for slower craft that could be built relatively soon, such as one powered by nuclear-pulse rockets, even if the journey takes several centuries.

In that case, however, the interstellar trip takes on another dimension. Rather than being simply a utilitarian transport vessel for a small band of explorers, the ship will have to function as a mobile cradle of civilization, equipped to handle births and deaths and to educate and socialize many generations of earthlings who will never have seen their planet of origin.

Some of the technology to build such an ark is straightforward enough. Both the United States and the Soviet Union have already demonstrated techniques for creating space habitats. In 1973, NASA launched *Skylab,* a space station that housed three successive crews, each for a period of one to two months over the course of about a year. The Soviets, for their part, have built and maintained a series of orbiting space stations almost continually since 1971. Cosmonaut Yuri Romanenko set a record for living in space when he spent 327 days aboard the Soviet station *Mir* in 1987.

For the present, *Mir,* which the Soviets continue to occupy periodically to conduct various experiments, must be supplied by ferries from Earth. However, engineers have been working on developing agricultural methods and systems for recycling air, water, and wastes to make space habitats self-sufficient. The Soviets have had some success at growing lettuce, peas, and a species of plant called arabidopsis in orbit, and other experiments with insects, fish, rats, and monkeys have given scientists valuable insights into the effects of weightlessness on living organisms. Terrestrial experiments are under way to construct an enclosed ecosystem capable of sustaining a small group of humans, along with plants and animals, with nothing supplied from the outside except sunlight.

Obviously, maintaining the delicate balance of life aboard ship will be a major prerequisite for the long interstellar voyage; failure of the recycling system, for instance, would have fatal consequences. (The redundancies built into spacecraft design would make such an occurrence unlikely except in the event of a massive collision.) Probably more difficult to manage, though, will be the social aspects of the undertaking. Humankind has lived in small, isolated groups before, whether deep in the jungles of Africa, in the remote hills of Appalachia, or on coral atolls in the Pacific. Small bands of like-minded individuals have also tried to isolate themselves on purpose, in order to establish utopian colonies of one stripe or another. But there has never been such a drastic severing of ties with the rest of humanity as interstellar explorers will experience. Whoever organizes the expedition will be faced with trying to anticipate possibly profound evolutionary changes in psychology and social behavior.

A COMPATIBLE CREW

Clearly, the most painstaking care must be exercised in selecting the original crew. According to one estimate, simply to operate the ship, study the environment of the target planet, and found a colony would require crew members with expertise in nearly a hundred different fields of science and engineering. Given the projected duration of the voyage and the need to control population growth, the starting population may not number more than 200 persons. Those pioneers will need to possess not only a high degree of intellectual versatility but also strong tendencies toward cooperation rather than competition. They will have to be willing to share in work such as raising food and educating children, and adults will also need to be lifelong students, acquiring new skills and making advances in their own fields to pass on to their descendants.

Some sort of governing structure would be necessary, if only to allocate workers to keep the ship running and manage common endeavors such as education and healthcare. But—assuming a certain psychological profile for anyone willing to venture into the great unknown—it would need to strike a fine balance between being as authoritarian as, say, a military hierarchy and as thoroughly democratic as a New England town meeting.

The society would also need to maintain the group's singularity of purpose beyond the second or third generation. Some scientists have noted that a strong religious or quasi-religious underpinning tends to foster group unity and obedience to authority, but that approach brings its own drawbacks.

Chief among them, perhaps, is the potential tendency to stifle individuality, diversity, and creativity—qualities that will be vital during the long journey and essential to the survival of the colony once it reaches landfall.

OF METHUSELAHS AND CYBORGS

The difficulty inherent in handing down the mission from generation to generation on a journey of many centuries would vanish if, through techniques of bioengineering, the original crew could somehow live long enough to found the colony at the other end. Some biologists speculate that it may be possible to develop a so-called Methuselah enzyme, named for the patriarch who, according to biblical references, lived for 969 years. In theory, such a chemical would retard all of the processes that age the body. Another approach to longer life would be to alter the very genes that control aging.

If no workable way could be found to slow or eliminate the ravages of age, a crew member could still be kept alive by substituting mechanical devices for the defective ones. Medical science is already capable of manufacturing highly sophisticated artificial limbs, and has made a beginning toward replacements for the kidneys, heart, skin, and blood. Over time, the crew could be transformed into cyborgs, beings who are part human, part machine. The ultimate cyborg might be a computer that, in effect, houses an individual brain, whose neural patterns—including knowledge and emotions—could be re-created in electronic form.

All of these scenarios for extending the lives of crew members presuppose that the individuals remain awake and functioning for the duration of the journey. Another possibility, again with an eye toward minimizing social disorder, is that the crew could spend most or even all of the trip in a state of suspended animation. On arrival, they would scarcely have aged and would have years of productive life ahead.

Several phenomena currently under investigation for their inherent scientific interest may radically affect the experience of future spacefarers. Hibernation is one such process. Every winter, certain species of mammals do not eat, drink, or excrete waste products for periods ranging from a few weeks to several months. Body temperature falls and all metabolic processes slow down. The temperature of one type of ground squirrel, for example, drops to thirty-four degrees Fahrenheit, and its heart rate slows from 350 beats per minute to 2. Every two or three weeks, the squirrel wakes up for perhaps a day, during which time its temperature and bodily functions return to normal, before returning to hibernation. The black bear, for its part, is unique in that it does not undergo such periods of arousal for the entire duration of the winter. The wastes produced by its metabolism are chemically broken down within the body and reformulated into proteins for nourishment.

In 1987, physiologist David Bruce of Wheaton College in Illinois reported having found evidence of a single compound that apparently circulates in the blood of black bears—as well as ground squirrels, woodchucks, and some bats—that triggers the metabolic changes associated with hibernation. Sim-

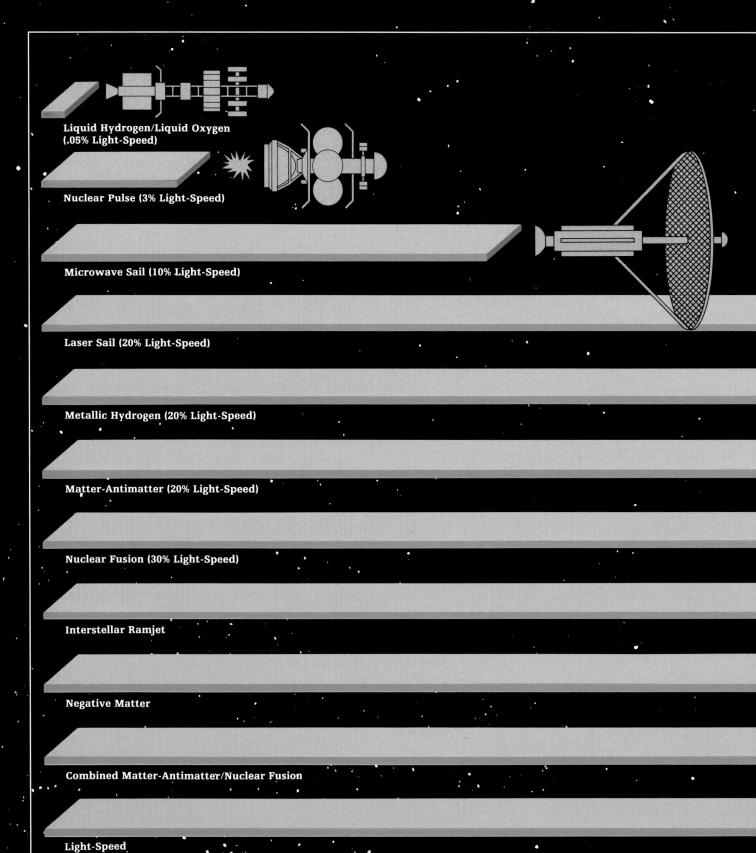

Liquid Hydrogen/Liquid Oxygen
(.05% Light-Speed)

Nuclear Pulse (3% Light-Speed)

Microwave Sail (10% Light-Speed)

Laser Sail (20% Light-Speed)

Metallic Hydrogen (20% Light-Speed)

Matter-Antimatter (20% Light-Speed)

Nuclear Fusion (30% Light-Speed)

Interstellar Ramjet

Negative Matter

Combined Matter-Antimatter/Nuclear Fusion

Light-Speed

A Futuristic Fleet

If earthly ships are ever to ply the incomprehensible reaches of the interstellar ocean, scientists must devise propulsion systems capable of making those voyages humanly feasible. The graph below and on the following pages compares the potential speeds of spacecraft equipped with various methods of propulsion. For consistency in making the comparisons, all the ships carry a 500,000-pound payload consisting of a human crew, their supplies, and other mission equipment. All are also built in Earth orbit to avoid factoring in extra fuel to escape the planet's gravity. Because the weight of the propulsion system and fuel varies with the type of system, the speeds noted for each propulsion method are calculated according to an optimum mass-to-fuel ratio.

A few of these systems and their propellants currently exist, and some, such as nuclear fusion and metallic hydrogen, are in early exploratory stages. Others—notably tachyons, which travel faster than light itself—are purely theoretical.

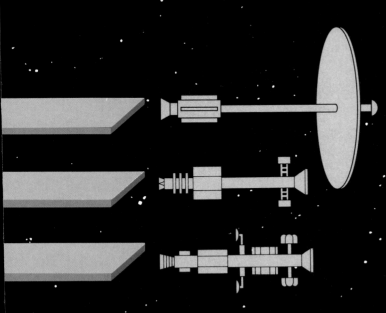

ilar clues had come to light several years earlier. In the early 1980s, zoologist Peter Oeltgen of the University of Kentucky experimented with removing blood plasma from hibernating woodchucks and injecting it into rhesus monkeys, a species that does not normally hibernate. The monkeys' heart rates slowed dramatically and they lost their appetites for up to three weeks without any significant loss of weight.

In addition to examining mammalian hibernation, scientists are intrigued by species that enter a more profound state of suspended animation, in which their body temperature falls below freezing. For most animals, such temperatures are fatal, but some fish are protected by a substance that acts as an antifreeze, dramatically lowering the point at which the body's water turns to ice. Such fish can survive temperatures down to 28.6 degrees Fahrenheit.

Space visionaries speculate that if the compounds that make up the piscine antifreeze or the mammalian hibernation trigger can be isolated and their structure determined, versions suitable for humans could be manufactured and administered to a flight crew on a lengthy journey. Granted, the nature of the interstellar vessel would undergo profound change. Instead of a lively

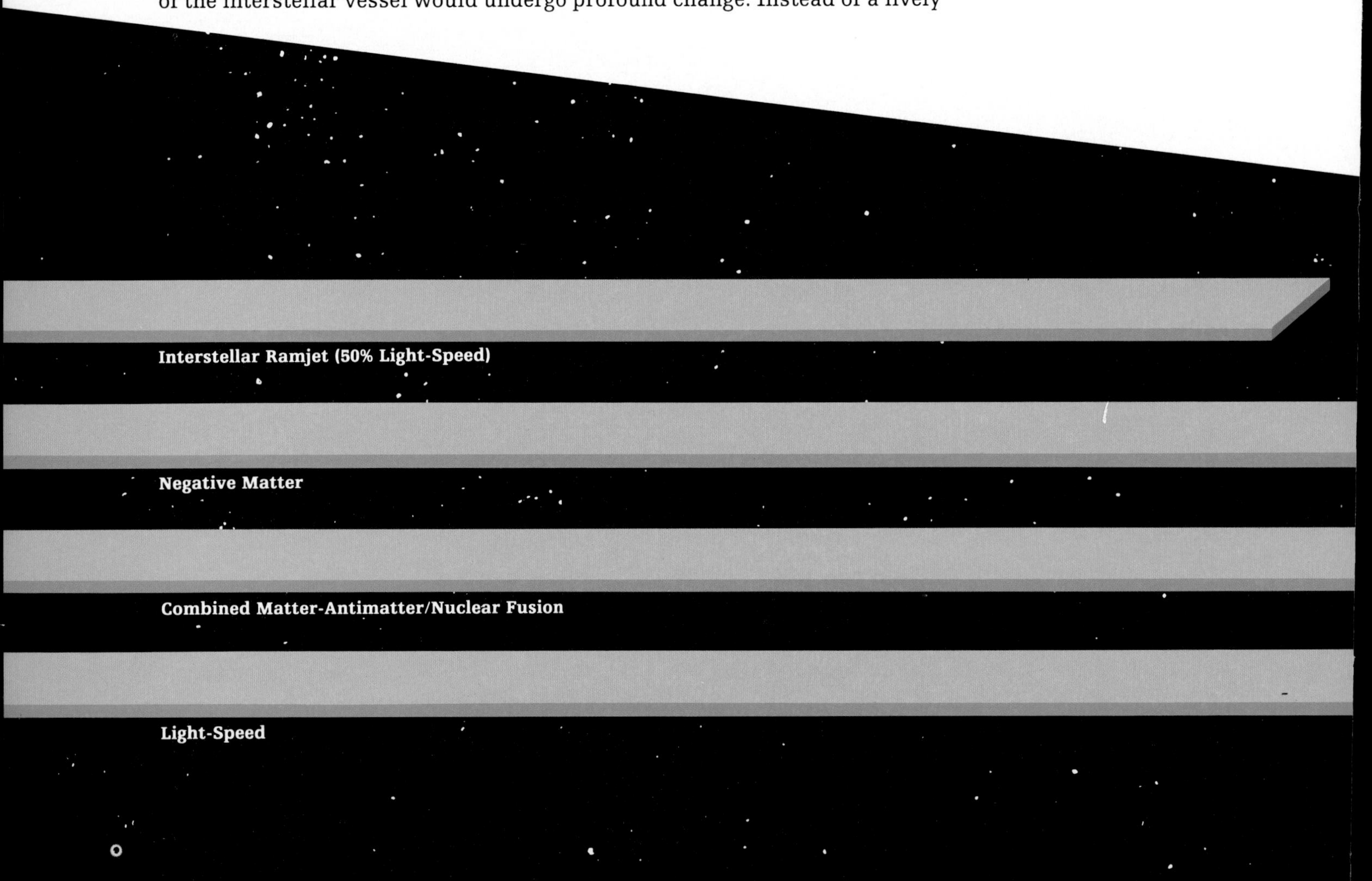

Interstellar Ramjet (50% Light-Speed)

Negative Matter

Combined Matter-Antimatter/Nuclear Fusion

Light-Speed

space village, the ship would become an interstellar refrigeration car. Crew members might take turns awakening every other decade or so to tend their immobile colleagues and to carry out necessary maintenance and repairs.

Somewhere down the road to the stars, space travelers may be able to trade in their lumbering arks—and many of the survival issues associated with them—for models equipped with interstellar ramjets or other engines that can accelerate to 50 percent of light-speed or more. Other factors now enter the equation, however, as Einstein discovered when he was working out his general and special theories of relativity. Some relativistic effects, as these phenomena are known, can work to a starfarer's advantage; others make the enterprise an even trickier proposition than it already is.

Einstein showed that as an accelerating object nears the speed of light, time slows down on board relative to a stationary observer. For example, voyagers aboard an interstellar ramjet accelerating to 50 percent light-speed would

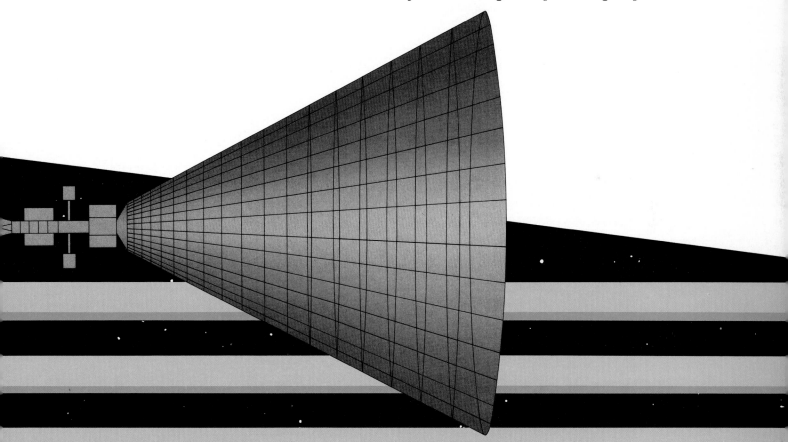

experience nine-tenths of a year for each year that passed on Earth. The spacefarers would thus age more slowly than their earthbound counterparts.

Another thing that changes at relativistic speeds is the mass of an object, which increases constantly as the object accelerates toward the speed of light. This has adverse consequences for a spacecraft attempting to approach light-speed because it is caught in a vicious circle, needing ever more power to propel its ever-growing mass. To reach the speed of light itself would require the impossible—infinite power. For practical purposes, then, an interstellar ship would probably never seek to go faster than 87 percent of light-speed; beyond that, the rate of mass increase jumps off the charts.

Finally, conventional navigating techniques would have to be abandoned on a ship moving at relativistic speeds because of the effect of movement on the wavelengths of light. Due to a phenomenon known as the Doppler effect, light waves compress toward the shorter, higher-frequency, or blue, end of the electromagnetic spectrum as their source approaches an observer and lengthen toward the lower-frequency, or red, end as the source recedes. From a relativistic spaceship, then, stars would appear to change color as the ship moved toward them, and then change color a second time after the ship had zoomed past *(pages 125-133)*. Because stars are cataloged in part by color, the ship's navigational system would need to take the Doppler shift into account. To further complicate the task, the stars not only change color but seem to change position as well, shifting forward until they are crowded into a tight cone directly in front of the ship.

THE CUTTING EDGE OF PHYSICS

Difficult as it may be to imagine an earthling navigator, except in a novel, ever having to wrestle with the outlandish tricks played by relativity, physicists

Negative Matter (70% Light-Speed)

Combined Matter-Antimatter/Nuclear Fusion (99% Light-Speed)

Light-Speed

are prepared to explore any number of ideas that appear to fall outside the bounds of common sense. Indeed, the premise of much of the work attempted by theoretical physicists may be summed up in the saying, "Anything which is not prohibited is compulsory."

Among the more exotic concepts recently entertained as a possibility for powering vessels that can bridge the vast distances between stars is negative matter. Positive matter encompasses both ordinary matter and antimatter; negative matter is something else again.

Some three decades ago, Hermann Bondi, an Austrian-born physicist who, along with colleague Thomas Gold and British astronomer Fred Hoyle, championed the so-called Steady State theory of the universe in the early 1950s, wrung the idea of negative matter out of Einstein's equations for general relativity theory. The theory, whose focus is gravity, neither mentions nor rules out negative matter. In a paper published in 1957 in the thoroughly respectable journal *Reviews of Modern Physics,* Bondi asserted that if negative matter existed, it would, not surprisingly, behave very oddly. For one thing, it would exert what is now called negative gravity, repelling all forms of matter, both positive and negative. Positive matter, on the other hand, attracts all forms of matter, negative and positive alike. In carrying the idea to its logical conclusion, Bondi described an interesting scenario: If a body of positive mass and a body of negative mass were separated by just the right distance, they would move off with uniform acceleration. The positive mass would lead the way because it would be constantly attracting the negative mass, which in turn would be constantly pushing the positive mass away in the same direction that the negative matter is being pulled.

Thirty-one years later, futurist Robert Forward expanded on Bondi's notion. In 1988, he began to consider how negative matter, assuming its existence, could be employed to drive a starship. For starters, the mass of negative matter would have to equal the (positive) mass of the ship. Next, because neither the negative matter nor the ship would be dense enough to produce significant gravitational forces, the two masses would be given opposite

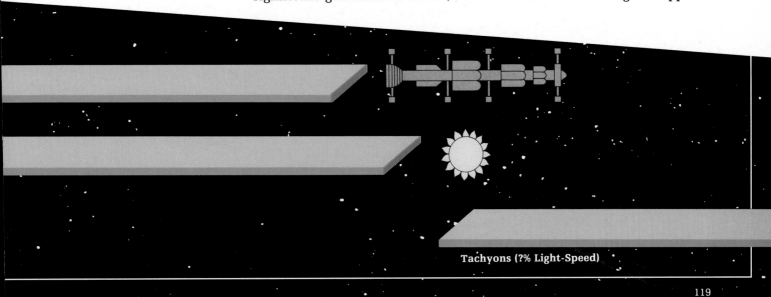

Tachyons (?% Light-Speed)

119

electric charges and linked by means of electrostatic forces to produce a configuration that would yield constant acceleration. To accelerate, the negative mass would be positioned just behind the ship's stern; to decelerate, it would be shifted to the front end of the ship. By one calculation, charged negative matter could accelerate the ship to 70 percent of light-speed, slashing the travel time to Epsilon Eridani to 12.5 ship years (17 years on Earth).

Several physicists besides Forward and Bondi have concluded that Einstein's theory does not exclude the possibility of negative matter, but finding a supply of it is problematic, to say the least. An undaunted Robert Forward has speculated on one likely place to look: In the last decade or so, a number of astronomical surveys have turned up evidence that galaxies are arranged in bubblelike structures. In effect, the galaxies seem to describe a spherical shell around enormous spaces devoid of stellar systems or any other detectable matter. Forward suggests that the voids may contain negative matter whose perverse gravity has pushed the galaxies away from it in all directions.

CROSSING THE GALAXY

Charged negative matter propulsion would be eminently useful for relatively short jaunts to the Sun's nearest neighbors, and it would also be suitable for even longer trips with multigenerational crews. The system's chief advantage is that the fuel would never be used up, assuming an adequate amount of negative matter could be found in the first place. For a truly long interstellar reach—a 60,000-light-year trip to the other side of the galaxy, for example—another form of transportation is needed before the trip becomes feasible even for many generations. Obviously, even in a light-speed vessel the journey is impossibly long. A visionary spacefarer's dream ship thus would not hover just below light-speed but would jump past it. In this way, the ship would not, strictly speaking, be defying the law that says nothing can travel at the speed of light except light itself.

Every type of subluminal propulsion method and the matter involved in it would, of course, be ruled out, since the phenomenon of increasing mass that Einstein described would prevent it from reaching light-speed. But if some kind of matter moved at superluminal speed from the moment of its creation, it would not be in violation of the law. Theoretically, such a particle would be capable of infinite speed, and in fact—in the reverse of Einstein's dictum— to decelerate it to the speed of light would require infinite energy.

Physicists call this theoretical particle the tachyon, from the Greek word, *tachys,* for swift. So far it remains no more than an imaginary mathematical construct, but one that has been explored for more than two decades. Indeed, particle physicists have searched for evidence of tachyons in the debris resulting from high-energy bombardments in particle accelerators. And even though they have yet to find such evidence, they have also not discovered anything that disproves their existence.

Of course, even if tachyons materialize, scientists would still face the vexing problem of crossing the Einsteinian barrier in order to take advantage of them for spaceflight. The theory of quantum mechanics, which describes the frequently counterintuitive behavior of the subatomic particles that make up ordinary matter, might help, however. One interesting idea is based on the fact that a subatomic particle, when given a jolt of energy, suddenly behaves more like a wave of energy than a particle of matter, and executes a "quantum jump"—suddenly appearing in a new location without actually traversing the space in between. Such a maneuver is counter to common sense, but particle physicists commonly witness such events in their experiments, notably in the behavior of electrons and photons.

Thus, in theory, if an entire spaceship could manage a quantum jump, it might be able to pass from, say, 99 percent of light-speed to a speed just slightly higher than that of light, without having traveled at light-speed itself. One way to make the jump might be with a ship propelled by a combination of systems—nuclear fusion plus matter-antimatter annihilation, for example, which would get the ship to 95 percent of light-speed. A special matter-antimatter booster could then be used to achieve 99 percent of light-speed. The acceleration to 99 percent light-speed need not be sustained for long; a pulsed magnetic field could push the ship past the light-speed barrier. Then, in a fraction of an instant, a tachyon drive could kick in, accelerating the ship to an incredible 300 times the speed of light. A trip to Epsilon Eridani would take a mere 6.5 ship-years, including time for acceleration and deceleration; one to the other side of the galaxy would take only 200 years. (The scheme has one ironic consequence. Instead of time slowing down aboard ship—relative to Earth—as it does at subluminal speeds, it speeds up; thus the tachyon travelers are aging slightly faster than their cousins on Earth.)

THE ULTIMATE SHORTCUT

Tachyons may not be the only route to faster-than-light travel. Tunnels might be created through space to serve as shortcuts between two widely separated points. Again, the seed of this incredible notion may be found in Einstein's general relativity theory. Einstein demonstrated that the four-dimensional fabric of space-time is warped by the presence of mass and that gravitational attraction is simply the effect of objects following the curvature of space-time. Conceivably, two objects could be so massive that the resulting warping creates a kind of connection—dubbed a wormhole—between them. The only objects that could have this effect, of course, are black holes, the almost infinitely dense remnants of massive stars. So intense is the gravity of a black hole that no objects, not even light, can escape its maw. But if black holes are the gateways to wormholes, how can a ship survive the unimaginable gravitational forces to which it would be subjected? And in any case, mathematical calculations show that the diameter of the wormhole is smaller than a subatomic particle, and the hole would pinch closed an instant after it formed.

Physicist Kip Thorne of the California Institute of Technology, taking advantage of the rule that everything not forbidden is compulsory, has derived from quantum theory an imaginary substance to prop a wormhole open.

Even faster than a tachyon-drive ship for interstellar travel would be a wormhole. Predicted by extrapolations of Einstein's general theory of relativity, wormholes would be the result of the tremendous warping of space-time caused by supermassive black holes. If a way could be found to traverse the black holes safely, a ship might take an instant shortcut several billion light-years across the cosmos.

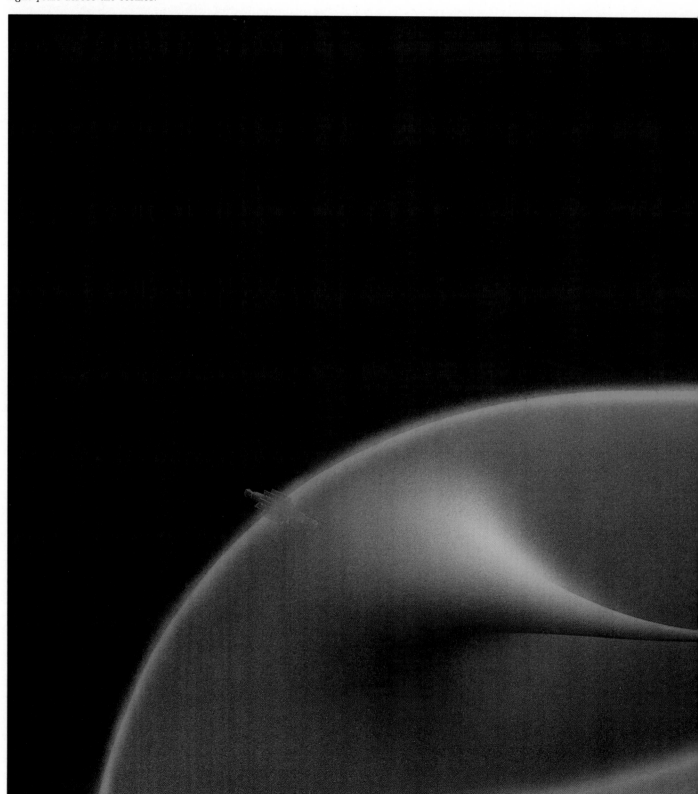

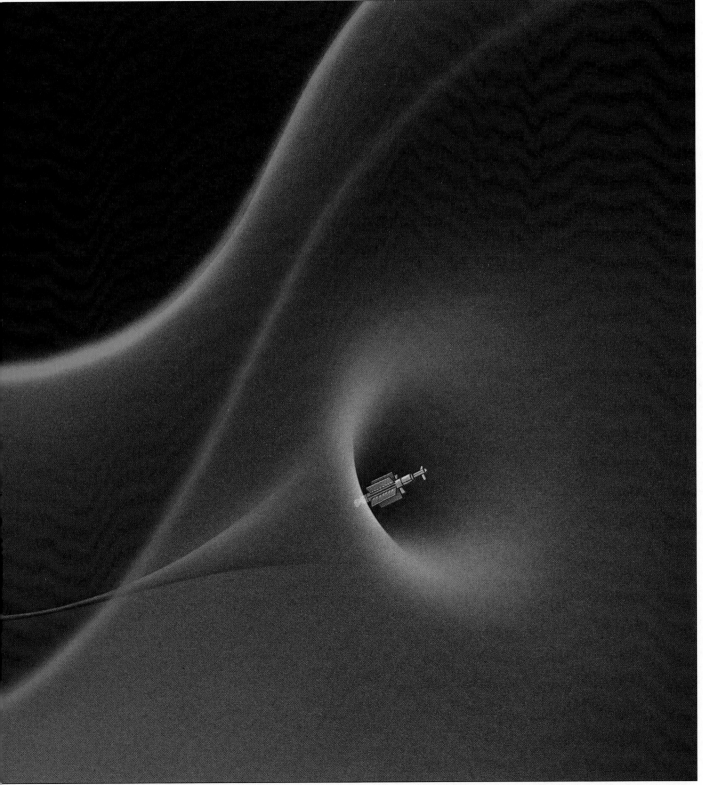

Known as "exotic matter," it has an extraordinarily high density, greater than that of the wormhole. The exotic matter exerts an outward pressure—comparable to that at the heart of a neutron star—strong enough to prevent the wormhole's collapse, allowing a spaceship to negotiate the tunnel safely.

Another way to take advantage of space warping is being explored by a number of researchers and engineers. Instead of risking the dangers and difficulties of natural black holes and wormholes, they envision a spacecraft that could weave a black hole and a wormhole around itself to transport it safely to a distant location in space-time. To generate black holes and wormholes on command would require a considerably improved understanding of the quantum and cosmological interactions that underlie the apparently smooth fabric of space-time, however.

Researchers looking into the creation of wormholes postulate that the universe within instants of the Big Bang had ten dimensions or more and a single, unified force. Almost immediately, however, the ten dimensions collapsed into the four that characterize the space-time universe we live in. At the same time, the single force differentiated into the four separate forces: gravity, electromagnetism, the weak nuclear force, and the strong nuclear force. But if a ten-dimension, one-force universe once existed, these thinkers suggest, it may still exist in a latent form, a so-far undetected foundation of the observable universe. Some scientists speculate that there may be a way to modulate electromagnetic energy patterns so as to interact with this underlying structure. The interaction might produce unexpectedly large effects, much as a glass can be shattered by sound at a particular, resonant frequency. By generating these electromagnetic patterns in or around a spacecraft, these scientists suggest, a hyperspatial connection—a wormhole—could be created between the spacecraft and a distant point in space-time.

Just as intercontinental air travel and the invention of radio and television turned Earth into a global village, portable wormholes would make trips to Epsilon Eridani—or the Andromeda galaxy—as feasible as flying to the Bahamas. The idea may be far-fetched today, but bold ideas are the inspiration for the imaginative technological breakthroughs that ultimately transform dreams into reality.

Toward the end of the nineteenth century, leading physicists were reportedly advising their students to go into other fields, because all the great discoveries in physics had already been made. The only tasks these eminences could envision were on the order of refining existing measurements by a decimal point or two. Within just a few decades, however, the supposedly quiescent field was utterly transformed by the discovery of subatomic particles and the advent of relativity theory and quantum mechanics.

Scientists and engineers of today, much less complacent than their forebears, assume that more—and more profound—changes lie in store. Whether the concepts of visionaries like Forward and Bussard will prove practical in the future or will have served only to keep the dream of starfaring alive, their work will stand among humanity's first practical steps to the stars.

Any hope that humans can travel between the stars in the course of a single lifetime depends on devising ships that will reach velocities approaching the speed of light—186,283 miles per second, or 5 trillion 875 billion miles per year. In visual terms, the experience of accelerating to such improbable speeds would be bizarre, a kind of optical magic show orchestrated by the theory of relativity. As spelled out in Einstein's equations, observers traveling at relativistic speeds would perceive a significant warping of space. For interstellar wayfarers streaking across the universe at 80 percent of light-speed, stars positioned astern would seem to migrate forward in the sky, growing collectively brighter as they crowded into a smaller area. They would also change color because of the stretching or compressing of their light's wavelengths by the effect known as Doppler shifting. By the time the velocity of the ship exceeds 97 percent of the speed of light, most of the visible universe would appear to have bunched ahead of the ship; only blackness would be seen behind.

The English physicist Eric Sheldon has explored the strange illusions of light-speed travel in a computer model dubbed STELLA, sampled on the following pages. Sheldon's program assumes the constant acceleration of a starship at a comfortable 1g, equivalent to the force of gravity on Earth. This steady gain, amounting to 79,000 miles per hour every hour, is more than ample for interstellar travel. Indeed, this incredible acceleration would allow a ship and its mortal passengers to traverse 18 billion light-years—the span of the entire visible universe—in just twenty-three years.

Stellar Illusions

The universe as observed by near light-speed voyagers has been likened to an umbrella being blown inside out. Because the paths of light traveling from celestial objects to the eyes of the wayfarers are dramatically altered by the ship's high-velocity passage through the cosmos, space itself seems reshaped. As a result, stars and galaxies that once surrounded the onrushing craft seem to slide forward along the vessel's course, drawing closer all the while. The diagram below depicts the apparent movement of a single star as observed by astronauts in a ship *(blue, upper right)* that has accelerated to within a fraction of light-speed.

Even as a star appears to change position in space by relativistic warping, all of the frequencies of its radiation seem to shift as the shipboard observers move away from the light source or toward it. The color of stars ahead of the ship shifts toward the blue end of the visible spectrum, then continues shifting into the invisible ultraviolet range and beyond. For stars located astern at the start of the voyage, the Doppler shifting is more complicated. At first, their light shifts toward the red end of the spectrum by the stretching of wavelengths as the ship speeds away. Then, as space warping alters their light paths to a forward position, the wavelengths are compressed and blueshifted. The diagrams at far right show relativistic effects as observed through the front and rear windows of an accelerating ship; in each case, the case history of a single star is chronicled. The only stars that would not undergo the apparent relocation shown in the diagrams would be those located precisely along the ship's line of travel, either ahead or behind. However, they would gradually fade from view because of Doppler shifting—into the infrared for a star directly astern, and into the ultraviolet for one directly ahead.

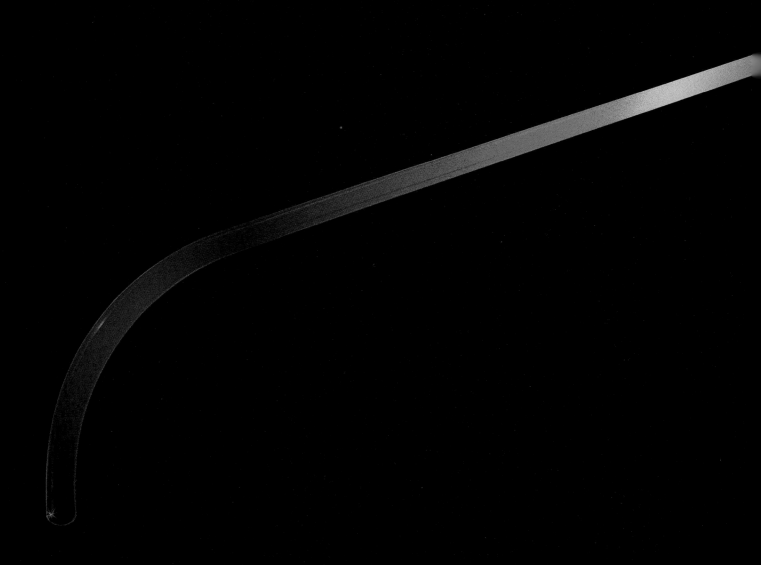

Depicted below are the apparent movement and frequency shifts of an ultraviolet star located 178 degrees away from the direction in which an interstellar vessel is traveling. At the start of the journey *(below, center)*, with the ship at rest, the star is 3,000 light-years to the rear, and its radiation is invisible. As the ship approaches light-speed and the star seems to move around to the side, its radiation enters the visible range, first as violet light, then blue, green, white, yellow, orange, and red. When the radiation shifts into the infrared, the star disappears again, only to reappear as its apparent motion carries it in front of the spacecraft and the frequency of its radiation compresses from red back toward violet; at this point, it appears to be only 104 light-years away. Finally, when dead ahead of the accelerating ship, which has traveled 2,164 light-years *(far right)*, the star vanishes into the ultraviolet.

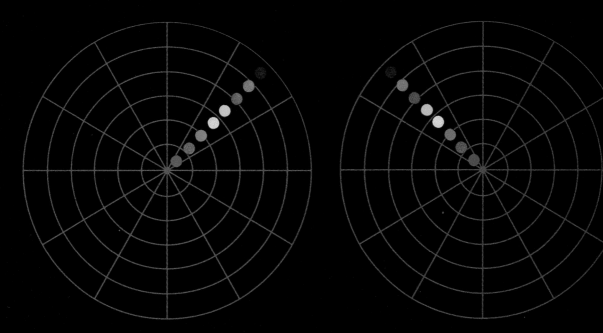

Views from the rear and front windows of a starship—each taking in half the sky—trace the illusory changes displayed by single stars as a ship at rest accelerates to 99.9 percent of light-speed. A star almost directly behind the ship *(below, left)* seems to move outward from the central area of the window to the side. As it moves, the star also changes color, shifting from the extreme blue toward the red end of the spectrum. A star in the forward field of view *(below, right)* undergoes an opposite color shift as it moves inward from the side.

A Cosmic Kaleidoscope

For voyagers crossing the starry universe, the relativistic movements and color shifts observed in the sky would play like a slow-motion kaleidoscope. At right and on pages 130 and 131, the complexity of the light show is suggested in a series of 180-degree forward views from within a ship accelerating to near light-speed, headed toward Polaris, the North Star. The initial view through the ship's rear window *(below, left)* shows constellations familiar in the Southern Hemisphere; the northern sky appears in the forward view *(below, right)*. The stars' colors, barely detectable from Earth, have been exaggerated for clarity. Beneath the window vistas are representations of the travelers' sphere of vision, with the ship at the center. As they accelerate, the warping of space causes a cone of darkness to grow astern, pressing the visible universe into an ever-smaller portion of the sphere.

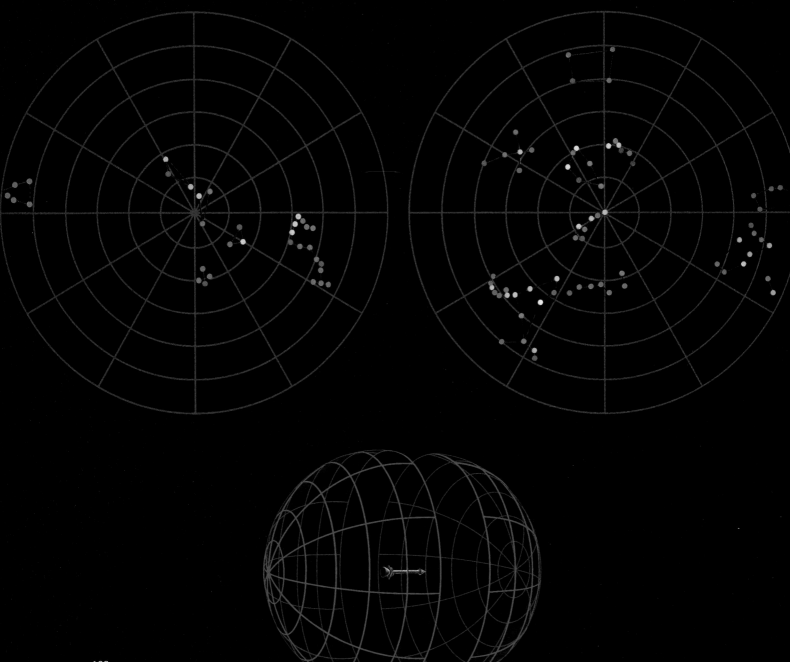

50 percent of light-speed. Having gone 88 billion miles and reached a velocity of about 93,142 miles per second, the starbound travelers begin to notice some relativistic changes. Constellations in front of the ship are still recognizable but have migrated slightly inward. Some stars that were originally invisible have shifted from infrared to visible red, orange, and yellow; other stars have also changed color and now appear green or blue. With the apparent front-ward movement of stars, the ship's spherical field of vision has been slightly diminished: A cone of darkness opens to the rear *(bottom)*.

80 percent of light-speed. Having accelerated at 1g for 1.06 ship years, the craft is 3.8 trillion miles from Earth and moving at 149,026 miles per second. Stars—most shifted into the blue-violet end of the spectrum—are bunching up in the forward window, and some constellations have grown harder to distinguish. Scorpius *(lower left quadrant)*, which started out behind the ship and had moved partly into the forward view at half light-speed, is now entirely within the front field of vision. The cone of darkness has enlarged considerably behind the ship.

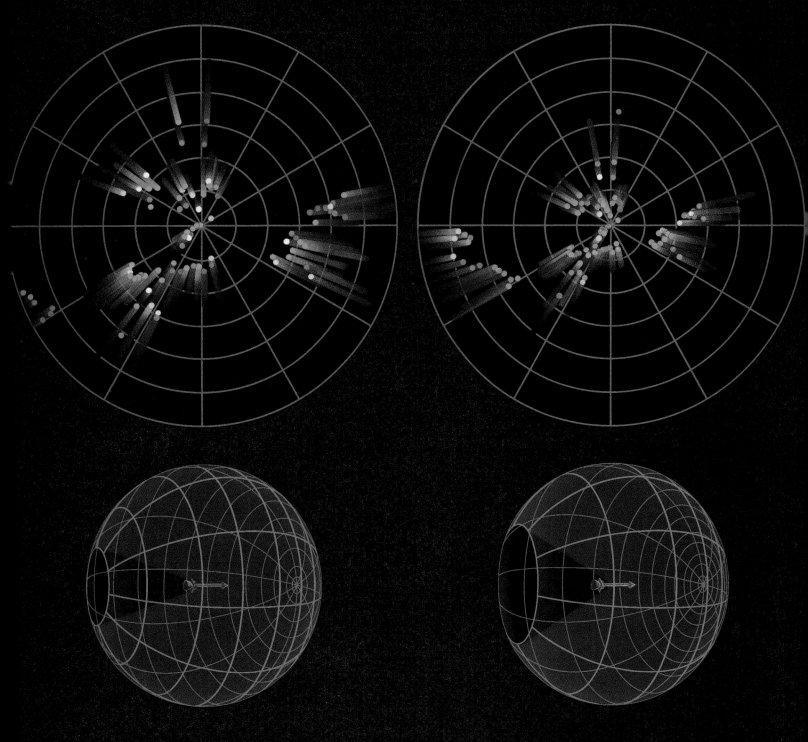

90 percent of light-speed. After 1.4 ship years (two years on Earth), the star ship is hurtling along at 167,655 miles per second, and stellar congestion is more pronounced. All the stars originally in the front half of the ship's sphere of vision now appear within 30 degrees of the center of the forward window. Spatial distortion has brought three Southern Cross stars—now green—from the rear into view at the bottom of the forward window. With blueshifting, some once-visible stars have vanished into the ultraviolet. The growing cone of darkness now blots out a significant portion of sky behind the ship.

98.5 percent of light-speed. The star-bound vessel, having traveled a distance of 4.6 light-years in 2.4 ship years (5.5 Earth years), is now moving at 183,489 miles per second. Most visible stars have shifted all the way to the blue-violet end of the spectrum. Of the original front view constellations, Ursa Major (the Big Dipper) has disappeared, and the remains of the rest are crammed together in the center of the front window. The stars farthest from the center of the window are from constellations that are actually within 30 degrees of the south celestial pole, originally visible out the ship's rear window—where there is now nothing to be seen *(bottom)*.

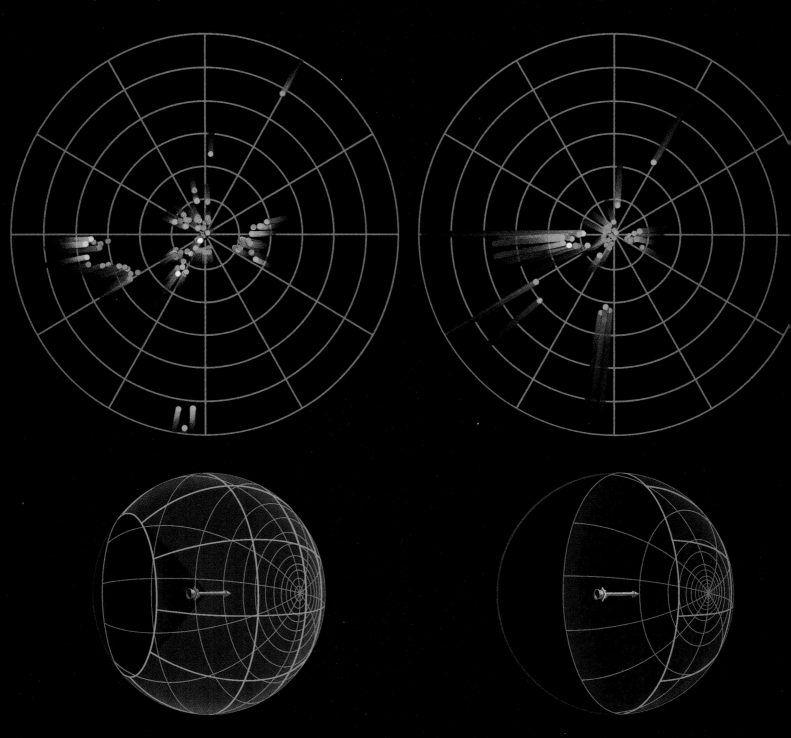

99.9 percent of light-speed. The starship has now traveled a distance of 20.7 light-years in 3.7 ship years (21.6 Earth years). Of the 101 stars shown in the original front and rear views, only 14 are still visible, and they are all clustered within 15 degrees of the center of the forward field of view. Most stars have shifted into the ultraviolet, with one speck of red provided by a star from the southern constellation Hydrus that has emerged from the invisible wavelengths of the far infrared. The starship's original sphere of vision has shrunk to a cone just 58 degrees across; all else is black.

99.999 percent of light-speed. The starship has traveled a distance of 216 light-years in 5.9 ship years (217 Earth years). At this point in the journey, the travelers' original sphere of vision has been almost completely blacked out; nothing appears outside a narrow six-degree cone in front of the ship. Most of the visible stars are those that started almost directly behind the ship. These few stragglers will soon shift into invisibility as well, leaving the universe in total darkness.

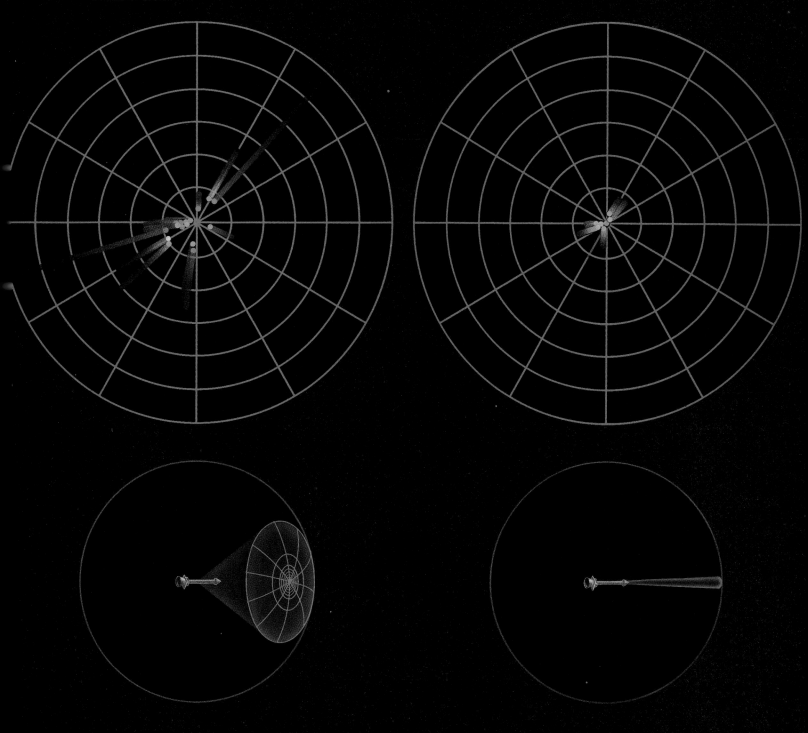

Light-Speed's Optical Implosion

If a starship could accelerate almost instantaneously to near light-speed, the entire visible universe, flashing through a rainbow of colors, would rush to a point directly ahead of the vessel *(left)*—a revved-up, science-fiction-style version of the relativistic effects traced on the preceding pages. At 99.99999 percent of light-speed, all stars will have Doppler shifted into invisibility, leaving the universe completely dark from a shipboard perspective.

This darkness will last just over a year for the voyagers, until the ship's velocity climbs to 99.999999 percent of light-speed. The protracted night will then be relieved only by Doppler shifting of the lowest-energy radiation in the universe, the fossil background radiation released in the primordial fireball of the Big Bang. Detected on Earth as weak microwaves coming from everywhere in the sky, this cosmic background radiation will be shifted into the visible portion of the electromagnetic spectrum by the ship's speed. All of the Big Bang afterglow, like all of the stars, will be concentrated in front of the ship by spatial warping. The interstellar travelers will see it dead ahead, appearing as a blue dot about one-tenth the brightness of the Sun and about one-sixtieth of a degree across.

Voyagers accelerating out from the Solar System at a constant rate of 1g will achieve cosmic dot velocity after traveling for nine ship years and traversing 6,846 light-years of interstellar space. So great is the time-warping effect of near light-speed travel that 6,847 years will have passed on Earth by then. The dot will remain visible as long as the ship continues at a pace close to the speed of light, vanishing again only when deceleration begins. Then all of the speed-induced illusions of the journey will be repeated in reverse as the interstellar travelers brake for their cosmic landfall.

GLOSSARY

Acceleration: a change in velocity. The term includes changes of direction and decreases as well as increases in speed.

Antimatter: matter made up of antiparticles identical in mass to matter particles, but opposite them in such properties as electrical charge.

Asteroid: any small, rocky, airless body that orbits a star. Three main kinds have been identified in the Solar System: C-type, or carbonaceous; S-type, or silicaceous; and M-type, or metallic. Asteroids that come near to Earth, and in some cases cross Earth's orbit, are called Earth approachers.

Astronaut: a pilot or passenger on a space flight. Soviet astronauts are called cosmonauts.

Atmosphere: a gaseous shell surrounding a planet or other body.

Atmospheric pressure: the weight of atmospheric gases on surfaces within a planet's atmosphere. For example, Earth's atmospheric pressure is about 14.7 pounds per square inch at sea level. Such pressure is also supplied artificially in spacecraft and in spacesuits.

Biosphere: the totality of a planet's living things and their habitats.

Black hole: theoretically, an extremely compact body with such great gravitational force that no radiation can escape from it. Proposed varieties include primordial, or mini, black holes, low-mass objects formed shortly after the beginning of the universe; stellar black holes, which form from the cores of very massive stars that have gone supernova; and supermassive black holes, equivalent to millions of stars in mass and located in the centers of galaxies.

Carbonaceous asteroid: an asteroid that contains carbon compounds.

Centrifugal force: the apparent outward force felt by a body rotating about an axis.

Centripetal force: the force that tends to keep moving matter on a curved path and pulled toward a central point.

Chlorofluorocarbons (CFCs): volatile, inert chemical compounds containing carbon, chlorine, fluorine, and hydrogen, often used as spray-can propellants and refrigerants. In Earth's upper atmosphere, the interaction of CFCs and ultraviolet radiation releases chlorine atoms in a process that accelerates the destruction of ozone, weakening the planet's shield against biologically dangerous radiation.

Chondrite: the most abundant subgroup of stony meteorites, characterized by the presence of small nodules of silicate.

Coriolis effect: the apparent deflection of an object's trajectory over the surface of a rotating body—a consequence of the body's rotation. The effect is seen, for example, in the spiral shapes taken by storms on Earth and in the atmospheres of other planets.

Cosmic ray: an atomic nucleus or other charged particle moving at close to the speed of light, thought to originate in supernovae and other violent celestial phenomena.

Cosmic string: according to theory, a type of massive one-dimensional object that formed during the early expansion of the universe.

Deceleration: negative acceleration; slowing.

Despinning: the process of slowing or halting the rotation of a body such as an asteroid or satellite.

Doppler shift: a change in the wavelength and frequency of sound or electromagnetic radiation, caused by the motion of the emitter, the observer, or both.

Electromagnetic energy (radiation): waves of electrical and magnetic energy that travel through space at the speed of light.

Equinox: one of two times a year when the Sun is exactly above the equator, and day and night are of equal length.

Escape velocity: the minimum speed needed for an object's momentum to carry it out of the gravitational pull of a massive object such as a planet or a moon.

Exobiology: the study of, and search for, organisms not native to Earth. No such organisms are yet known.

Exotic matter: theoretical particles invoked to explain certain observed effects of matter.

Gamma ray: the most energetic form of electromagnetic radiation, with the highest frequency and the shortest wavelength.

Geosynchronous: describing the orbit of a spacecraft or satellite that completes a circle every twenty-four hours, the same time as Earth requires to make one rotation; thus the object remains above one location on the ground. Geosynchronous orbits are established over the equator at an altitude of approximately 22,300 miles.

Gravity: the force responsible for the mutual attraction of separate masses. *See* Microgravity; Zero gravity.

Gravity wave: a theoretical perturbation in an object's gravitational field that would travel at the speed of light.

Gravity well: a local distortion in the fabric of space-time near a massive body; analogous to a well or depression in a two-dimensional sheet.

Greenhouse effect: a phenomenon in which radiation is selectively transmitted and absorbed by gases in an atmosphere, admitting incoming, short-wavelength solar radiation but blocking outgoing, long-wavelength infrared, thus trapping heat near the surface of a planet.

Greenhouse gas: an atmospheric constituent—such as carbon dioxide, nitrogen oxides, water vapor, chlorofluorocarbons, or methane—that warms the atmosphere by trapping outbound infrared radiation.

Helium: the second lightest chemical element and the second most abundant; produced in stars by the fusion of hydrogen.

Hominid: a primate of the genus *Homo,* including humans and their closely related ancestors.

Hydrogen: the lightest and most common element in the universe.

Ion: an atom that has gained or lost one or more electrons and has become electrically charged. In comparison, a neutral atom has an equal number of electrons and protons, giving it a zero net electrical charge.

Isotope: one of two or more forms of a chemical element that have the same number of protons but a different number of neutrons in the nucleus.

Laser: shortened from light amplification by stimulated emission of radiation; a device that produces a narrow beam of high-intensity monochromatic radiation at infrared, optical, or shorter wavelengths.

Laser sail: a proposed spacecraft propelled by the pressure of a focused beam of photons emitted by a laser against a sail-like surface.

Libration point: one of five locations between Earth and the Moon where the gravitational pull of the two bodies is balanced.

Light-speed: the speed light travels in a vacuum, 186,283 miles per second, or almost six trillion miles in a year.

Lunar orbit: the regular path of an object revolving around Earth's moon.

Magnetic sail (magsail): a spacecraft design proposing to use as propellant the interaction between the solar wind and an artificial magnetic field maintained around the spacecraft.

Mass driver: an electromagnetic catapult for launching objects into space, or for propelling spacecraft.

Meteor: a streak of light in the sky caused by the passage through Earth's atmosphere of a small rock or a piece of dust; also the luminous object itself.

Meteorite: the recovered fragment of a rocky or metallic body that has survived its transit through Earth's atmosphere. The weight of a meteorite may range from just a few ounces to nearly a hundred tons.

Microgravity: an environment—within an orbiting spacecraft, or on a small body such as an asteroid—of very weak gravitational forces. Microgravity conditions in space stations may allow experiments or manufacturing processes that are not possible on Earth.

Microwave: a radio wave of very high frequency and short wavelength.

Microwave sail: a proposed spacecraft built in the form of a large sail propelled by the pressure of microwaves against it.

Negative matter: a hypothetical form of matter having properties opposite those of positive matter; for example, separate masses of negative matter would repel rather than attract each other.

Neural network: in the field of artificial intelligence, a collection of computer processors linked together in such a way that individual processors can perform separate functions simultaneously, similar to the organization of neurons in the human brain.

Nuclear fission: a process that releases energy when heavyweight nuclei break down into lighter nuclei.

Nuclear fusion: a process that releases energy when lightweight atomic nuclei combine to form a heavier nucleus.

Nuclear-pulse rocket: a system designed to propel a spacecraft by repeated detonations of nuclear bombs. The explosions fire debris against a metal plate connected to the rear of the craft by massive shock absorbers; the ship is pushed forward each time the compressed shock absorbers rebound.

Ozone layer: a concentration of ozone—the highly unstable, three-atom form of oxygen—in Earth's upper atmosphere. The layer absorbs incoming solar ultraviolet rays that are harmful to life on the planet.

Particle: the smallest component of any class of matter; for example, the elementary particles within an atom (such as electrons, protons, and neutrons); or the smallest forms of solid matter in space (interplanetary and interstellar dust particles).

Photon: a packet of electromagnetic energy that behaves like a chargeless particle and travels at the speed of light.

Plasma: a gaslike conglomeration of charged particles that respond collectively to electrical currents and magnetic fields. Plasmas are considered a fourth state of matter along with solids, liquids, and gases.

Polar orbit: an orbit that crosses above the north and south poles of the body being orbited.

Probe: an automated, crewless spacecraft used to gather information or perform experiments in space or on extraterrestrial surfaces and transmit its findings back to Earth.

Propellant: a chemical or chemical mixture burned to create thrust for a rocket or spacecraft.

Quantum jump: a phenomenon of quantum mechanics in which a particle moves from one location to another without going through the intervening space.

Quantum mechanics: a mathematical description of the rules by which subatomic particles interact, decay, and form atomic or nuclear objects.

Radiation: energy in the form of electromagnetic waves or subatomic particles.

Ramjet: a jet engine that uses external propellant sources, such as gas collected from the interstellar medium, instead of carrying a supply of propellant.

Relativistic speed: motion at a significant fraction of the speed of light, in which changes in time, length, and mass become noticeable to observers who are stationary in relative terms.

Robot: a machine able to perform specific, usually programmed, functions, often used as a substitute for humans in inaccessible or dangerous environments.

Rocket: a missile or vehicle propelled by the combustion of fuel and a contained oxygen supply. The forward thrust of a rocket results when exhaust products are ejected from the tail.

Rotation: the turning of a celestial body around its axis.

Satellite: any body, natural or artificial, in orbit around a planet.

Seismometer: a device for measuring movements of the ground.

Siliceous asteroid: an asteroid rich in silicates.

Solar flare: an explosive release of charged particles and electromagnetic radiation from a small area on the surface of the Sun.

Solar sail: a proposed spacecraft for travel within the Solar System, propelled by the pressure of solar photons against vast expanses of ultrathin, highly reflective material.

Space station: a protected environment established in space, usually in orbit around a planet, as a base of operations for various space-related activities including manufacturing and scientific study.

Special theory of relativity: a theory postulating that observers in uniform motion cannot perceive their motion and that all observers in such motion obtain the same value for the speed of light. From these two principles the theory concludes that measures of distance, time, and mass will vary depending on the motion of an observer moving uniformly in relation to the thing being measured.

Spectrometer: an instrument that splits light or other electromagnetic radiation into its individual wavelengths, or spectrum, and records the results electronically.

Subatomic particle: *see* Particle.

Superluminal: appearing to travel faster than the speed of light.

Tachyon: a theoretical particle that only travels faster than the speed of light, without limit as to how fast it can go.

Terraforming: the theoretical process of altering a planet or satellite's natural state to create conditions more like those on Earth, rendering the planet suitable for human habitation.

Ultraviolet radiation: a band of electromagnetic radiation that has a higher frequency and shorter wavelength than visible blue light.

Wormhole: a hypothetical distortion in the fabric of space-time linking widely separated black holes; some theorists suggest wormholes could be used as tunnels for virtually instantaneous travel between widely separated points in the universe.

X-ray: a band of electromagnetic radiation intermediate in wavelength between ultraviolet radiation and gamma rays.
Zero gravity: a condition in which gravity appears to be absent. Zero gravity occurs when gravitational forces are balanced by the acceleration of a body in orbit or free fall.

BIBLIOGRAPHY

Books

Adelman, Saul J., and Benjamin Adelman. *Bound for the Stars.* Englewood Cliffs, N.J.: Prentice-Hall, 1981.

Allaby, Michael, and James Lovelock. *The Greening of Mars.* New York: St. Martin's/Marek, 1984.

Allen, Joseph P., and Russell Martin. *Entering Space: An Astronaut's Odyssey.* New York: Stewart, Tabori & Chang, 1984.

American Men & Women of Science (17th ed.). New York: R. R. Bowker, 1990.

Bainum, Peter M., et al. (eds.). *Advances in the Astronautical Sciences, Volume 62: Tethers in Space.* San Diego: American Astronautical Society, 1987.

Bellwood, Peter. *The Polynesians: Prehistory of an Island People.* London: Thames and Hudson, 1978.

Bernal, J. D. *The World, the Flesh and the Devil.* London: Jonathan Cape, 1970.

Binzel, Richard P., Tom Gehrels, and Mildred Shapley Matthews (eds.). *Asteroids II.* Tucson: University of Arizona Press, 1989.

Bova, Ben. *Welcome to Moonbase.* New York: Ballantine Books, 1987.

Breuer, Reinhard. *Contact with the Stars.* San Francisco: W. H. Freeman, 1982.

Clarke, Arthur C. *Arthur C. Clarke's July 20, 2019: Life in the 21st Century.* New York: MacMillan, 1986.

Close, Frank. *Apocalypse When? Cosmic Catastrophe and the Fate of the Universe.* New York: William Morrow, 1988.

Close, Frank, Michael Marten, and Christine Sutton. *The Particle Explosion.* New York: Oxford University Press, 1987.

Cole, Dandridge M., and Donald W. Cox. *Islands in Space: The Challenge of the Planetoids.* Philadelphia: Chilton Books, 1964.

Comets, Asteroids, and Meteorites (Voyage Through the Universe series). Alexandria, Va.: Time-Life Books, 1990.

Cosmic Mysteries (Voyage Through the Universe series). Alexandria, Va.: Time-Life Books, 1990.

Cunningham, Clifford J. *Introduction to Asteroids: The Next Frontier.* Richmond, Va.: Willmann-Bell, 1988.

Daintith, John, Sarah Mitchell, and Elizabeth Tootill (eds.). *A Biographical Encyclopedia of Scientists.* New York: Facts on File, 1981.

The Far Planets (Voyage Through the Universe series). Alexandria, Va.: Time-Life Books, 1988.

Ferris, Timothy. *Coming of Age in the Milky Way.* New York: William Morrow, 1988.

Finney, Ben R., et al. *Interstellar Migration and the Human Experience.* Berkeley: University of California Press, 1985.

Forward, Robert L. *Future Magic.* New York: Avon Books, 1988.

Forward, Robert L., and Joel Davis. *Mirror Matter: Pioneering Antimatter Physics.* New York: John Wiley & Sons, 1988.

Friedman, Louis. *Starsailing: Solar Sails and Interstellar Travel.* New York: John Wiley & Sons, 1988.

Gillispie, Charles Coulston (ed.). *Dictionary of Scientific Biography.* New York: Charles Scribner's Sons, 1978.

Goddard, Esther C., and G. Edward Pendray (eds.). *The Papers of Robert H. Goddard.* New York: McGraw-Hill, 1970.

Hadingham, Evan. *Early Man and the Cosmos.* New York: Walker, 1984.

Heppenheimer, T. A. *Colonies In Space.* New York: Warner Books, 1977.

James, Edward T., et al. (eds.). *Dictionary of American Biography.* New York: Charles Scribner's Sons, 1973.

Jastrow, Robert. *Journey to the Stars.* New York: Bantam Books, 1989.

Kaufman, William J., III:
The Cosmic Frontiers of General Relativity. Boston: Little, Brown, 1977.
Universe (2d ed.). New York: W. H. Freeman, 1987.

Kowal, Charles T. *Asteroids: Their Nature and Utilization.* New York: John Wiley & Sons, 1988.

Lambert, David, and the Diagram Group. *The Field Guide to Early Man.* New York: Facts on File, 1987.

Lehman, Milton. *This High Man: The Life of Robert H. Goddard.* New York: Farrar, Straus, 1963.

Lewin, Roger. *Human Evolution: An Illustrated Introduction.* New York: W. H. Freeman, 1984.

Lewis, John S., and Ruth A. Lewis. *Space Resources: Breaking the Bonds of Earth.* New York: Columbia University Press, 1987.

Lewis, Richard S. *Space in the 21st Century.* New York: Columbia University Press, 1990.

McAleer, Neil. *The Omni Space Almanac: A Complete Guide to the Space Age.* New York: Pharos Books, 1987.

McCormack, Percival D., and D. Stuart Nachtwey. "Radiation Exposure Issues." In *Space Physiology and Medicine* (2d ed.), edited by Arnauld E. Nicogossian. Philadelphia: Lea & Febiger, 1989.

McKay, Mary Fae, David S. McKay, and Michael B. Duke (eds.). *Space Resources.* Washington, D.C.: NASA, in press.

Macvey, John W. *Interstellar Travel: Past, Present, and Future.* New York: Stein and Day, 1977.

Mallove, Eugene F., and Gregory L. Matloff. *The Starflight Handbook: A Pioneer's Guide to Interstellar Travel.* New York: John Wiley & Sons, 1989.

Miles, Frank, and Nicholas Booth (eds.). *Race to Mars: The Mars Flight Atlas.* New York: Harper & Row, 1988.

Miller, Ron, and William K. Hartmann. *The Grand Tour: A Traveler's Guide to the Solar System.* New York:

Workman Publishing, 1981.

Minsky, Marvin (ed.). *Robotics*. Garden City, N.Y.: Anchor Press/Doubleday, 1985.

Motz, Lloyd, and Carol Nathanson. *The Constellations*. New York: Doubleday, 1988.

Murray, Bruce. *Journey into Space: The First Three Decades of Space Exploration*. New York: W. W. Norton, 1989.

The Near Planets (Voyage Through the Universe series). Alexandria, Va.: Time-Life Books, 1989.

Nicolson, Iain. *The Road to the Stars*. New York: New American Library, 1978.

Oberg, James E.:
Mission to Mars: Plans and Concepts for the First Manned Landing. Harrisburg, Pa.: Stackpole Books, 1982.
New Earths: Transforming Other Planets for Humanity. Harrisburg, Pa.: Stackpole Books, 1981.
The New Race for Space: The U.S. and Russia Leap to the Challenge for Unlimited Rewards. Harrisburg, Pa.: Stackpole Books, 1984.

Oberg, James E., and Alcestis R. Oberg. *Pioneering Space: Living on the Next Frontier*. New York: McGraw-Hill, 1986.

O'Leary, Brian:
The Fertile Stars. New York: Everest House, 1981.
The Making of an Ex-Astronaut. Boston: Houghton Mifflin, 1970.

O'Leary, Brian (ed.). *Space Industrialization: Volume I*. Boca Raton, Fla.: CRC Press, 1982.

O'Neill, Gerard K.:
The High Frontier: Human Colonies in Space. New York: William Morrow, 1977.
The High Frontier: Human Colonies in Space. Princeton, N.J.: Space Studies Institute Press, 1989.

Outbound (Voyage Through the Universe series). Alexandria, Va.: Time-Life Books, 1989.

Pagels, Heinz R. *The Dreams of Reason: The Computer and the Rise of the Sciences of Complexity*. New York: Simon and Schuster, 1988.

Pioneering the Space Frontier: The Report of the National Commission on Space. Toronto, Canada: Bantam Books, 1986.

Ponnamperuma, Cyril, and A. G. W. Cameron. *Interstellar Communication: Scientific Perspectives*. Boston: Houghton Mifflin, 1974.

Powers, Robert M. *The Coattails of God: The Ultimate Spaceflight—The Trip to the Stars*. New York: Warner Books, 1981.

Robinson, J. Hedley, and James Muirden. *Astronomy Data Book* (2d ed.). New York: John Wiley & Sons, 1979.

Rood, Robert T., and James S. Trefil. *Are We Alone? The Possibility of Extraterrestrial Civilizations*. New York: Charles Scribner's Sons, 1981.

Sagan, Carl. *Cosmos*. New York: Random House, 1980.

Spacefarers (Voyage Through the Universe series). Alexandria, Va.: Time-Life Books, 1990.

Stars (Voyage Through the Universe series). Alexandria, Va.: Time-Life Books, 1988.

Von Braun, Wernher, and Frederick I. Ordway III. *Space Travel* (3d ed.). New York: Thomas Y. Crowell, 1975.

Wilford, John Noble. *Mars Beckons*. New York: Alfred A. Knopf, 1990.

Yenne, Bill (ed.). *Interplanetary Spacecraft*. New York: Exeter Books, 1988.

Periodicals

Adair, Robert K. "A Flaw in a Universal Mirror." *Scientific American*, February 1988.

Aldrin, Buzz. "The Mars Transit System." *Air & Space*, October-November 1990.

Baer, John. "Artificial Intelligence." *The Futurist*, January-February 1988.

Bahcall, John N. "Neutrinos from the Sun." *Mercury*, March-April 1990.

Barnes, Brian M. "How Animals Survive the Big Chill." *Washington Post*, March 4, 1990.

Barnes-Svarney, Patricia. "Grabbing a Piece of the Rock." *Ad Astra*, October 1990.

Barton, William, and Michael Capobianco. "Harvesting the Near-Earthers." *Ad Astra*, November 1989.

Bartusiak, Marcia. "Sensing the Ripples in Space-Time." *Science '85*, April 1985.

Bekey, Ivan, and Paul A. Penzo. "Tether Propulsion." *Aerospace America*, July 1986.

Berry, Richard. "Neptune Revealed." *Astronomy*, December 1989.

Bilaniuk, Olexa-Myron, and E. C. George Sudarshan. "Particles Beyond the Light Barrier." *Physics Today*, May 1969.

Blass, W. Paul, and John Camp. "Society in Orbit." *Space World*, July 1988.

Bloomfield, Masse. "Sociology of an Interstellar Vehicle." *Journal of the British Interplanetary Society*, 1986, Vol. 39, pp. 116-120.

Bondi, H. "Negative Mass in General Relativity." *Reviews of Modern Physics*, 1957, Vol. 29, pp. 423-428.

Broad, William. "Ambitious Effort Aims to Find Gravity Waves." *New York Times*, February 27, 1990.

Burns, Jack O., et al. "Observatories on the Moon." *Scientific American*, March 1990.

Bussard, R. W. "Galactic Matter and Interstellar Flight." *Astronautica Acta*, 1960, Vol. 6, pp. 179-194.

Cassenti, Brice N. "A Comparison of Interstellar Propulsion Methods." *Journal of the British Interplanetary Society*, 1982, Vol. 35, pp. 116-124.

Chartrand, Mark. "Mars: Just the Facts." *Ad Astra*, May 1990.

Chrein, Lloyd. "Space." *Omni*, March 1990.

Cowen, R. "New Evidence of Budding Solar Systems." *Science News*, March 10, 1990.

Croswell, Ken. "Titan: Slumbering Giant." *Space World*, January 1988.

David, Leonard. "Notes from the Mars Underground." *Ad Astra*, August 1990.

Easterbrook, Gregg. "Are We Alone?" *The Atlantic*, August 1988.

Eberhart, Jonathan:
"From Earth to the Moon With Love." *Science News*, March 3, 1990.
"Space 1990: Launching a New Decade of Exploration." *Science News*, January 13, 1990.

Farrard, William. "On the Road Again." *Ad Astra*, May 1990.

Feinberg, G. "Possibility of Faster-Than-Light Particles." *The Physical Review*, July 25, 1967.

Flamstead, Sam. "Astronomy." *Discover*, January 1990.

Fogg, Martyn J.:

"The Creation of an Artificial Dense Martian Atmosphere: A Major Obstacle to the Terraforming of Mars." *Journal of the British Interplanetary Society*, 1989, Vol. 42, pp. 577-582.

"Stellifying Jupiter: A First Step to Terraforming the Galilean Satellites." *Journal of the British Interplanetary Society*, 1989, Vol. 42, pp. 587-592.

Forward, R. L.:

"Negative Matter Propulsion." *Journal of Propulsion and Power*, January-February 1990.

"Roundtrip Interstellar Travel Using Laser-Pushed Lightsails." *Journal of Spacecraft and Rockets*, 1984, Vol. 21, pp. 187-195.

"Starwisp: An Ultraviolet Interstellar Probe." *Journal of Spacecraft and Rockets*, 1985, Vol. 22, pp. 345-350.

Freedman, David:

"Beyond Einstein." *Discover*, February 1989.

"Cosmic Time Travel." *Discover*, June 1989.

Gaffey, Michael J., and Thomas B. McCord. "Mining Outer Space." *Technology Review*, June 1977.

"Galactic Center Antimatter Factory Found At Last?" *Sky & Telescope*, April 1990.

Garshnek, V. "Crucial Factor: Human." *Space Policy*, August 1989.

Glenn, Jerome C. "Conscious Technology." *Futurist*, September-October 1989.

Holt, Alan C. "Hydromagnetics and Future Propulsion Systems." *AIAA Student Journal*, Spring 1980.

Houtchens, C. J. "Artificial Gravity." *Final Frontier*, May-June 1989.

Ionasecu, Rodica, and Paul A. Penzo. "Space Tethers." *Spaceflight*, May 1988.

Jackson, Pat. "Next Floor: Space." *Ad Astra*, June 1990.

Kasting, James F., Owen B. Toon, and James B. Pollack. "How Climate Evolved on the Terrestrial Planets." *Scientific American*, February 1987.

Kolm, Henry. "An Electromagnetic 'Slingshot' for Space Propulsion." *Space World*, February 1978.

Lemonick, Michael D. "Back to Earth Unscathed." *Time*, February 1, 1988.

Lewis, Richard S. "Industrial Revolution Out in Space Promises Advances in Pharmaceutical and Metallurgical Technologies." *Smithsonian*, December 1977.

Lovelock, James E. "The Ecopoiesis of Daisy World." *Journal of the British Interplanetary Society*, 1989, Vol. 42, pp. 583-586.

McKay, Christopher P.:

"Martians Wanted: Dead or Alive!" *Ad Astra*, May 1990.

"Terraforming: Making an Earth of Mars." *The Planetary Report*, November-December 1987.

"Terraforming Mars." *Journal of the British Interplanetary Society*, 1982, Vol. 35, pp. 427-433.

McKay, Christopher P., and Robert H. Haynes. "Essay: Should We Implant Life on Mars?" *Scientific American*, November 1990.

McKay, Christopher P., and Carol R. Stoker. "The Early Environment and Its Evolution on Mars: Implications for Life." *Reviews of Geophysics*, May 1989.

"Magnetic Sailing Across Interstellar Space." *Ad Astra*, January 1990.

Mao, H. K., and R. J. Hemley. "Optical Studies of Hydrogen above 200 Gigapascals: Evidence for Metallization by Band Overlap." *Science*, June 23, 1989.

Merritt, J. I. "Pioneering the Space Frontier." *Princeton Alumni Weekly*, October 11, 1989.

Minsky, Marvin. "The Intelligence Transplant." *Discover*, October 1989.

Morrison, David. "Target Earth: It Will Happen." *Sky & Telescope*, March 1990.

Moskowitz, Saul. "Visual Aspects of Trans-Stellar Space Flight." *Sky & Telescope*, May 1967.

O'Leary, Brian:

"Asteroid Mining." *Astronomy*, November 1978.

"The Fertile Stars." *Quest/81*, May 1981.

"Mining the Apollo and Amor Asteroids." *Science*, July 22, 1977.

"Outward Bound." *Time*, January 27, 1961.

Penzo, P. A., and H. L. Mayer. "Tethers and Asteroids for Artificial Gravity Assist in the Solar System." *Journal of Spacecraft and Rockets*, January-February 1986.

Pool, Robert. "The Chase Continues for Metallic Hydrogen." *Science*, March 30, 1990.

Powers, Robert M. "The Ultimate Observatory." *Final Frontier*, March-April 1990.

Revkin, Andrew C. "Sleeping Beauties." *Discover*, April 1989.

Schechter, Bruce. "Searching for Gravity Waves with Interferometers." *Physics Today*, February 1986.

Sheldon, E., and R. H. Giles. "Celestial Views from Non-relativistic and Relativistic Interstellar Spacecraft." *Journal of the British Interplanetary Society*, 1983, Vol. 36, pp. 99-114.

Shiner, Linda. "300 Billion Watts, 24 Hours a Day." *Air & Space*, June-July 1990.

Sienko, Tanya. "Japan's Space Science." *Space*, March-April 1990.

Silk, Joseph. "Probing the Primeval Fireball." *Sky & Telescope*, June 1990.

SSI Update—High Frontier Newsletter, May-June 1990.

Stimets, R. W., and E. Sheldon. "The Celestial View from a Relativistic Starship." *Journal of the British Interplanetary Society*, 1981, Vol. 34, pp. 84-99.

Verrengia, Joseph B. "Americans Aim for Mars by the Year 2000." *Rocky Mountain News*, July 29, 1990.

Waldrop, M. Mitchell:

"NASA Flight Controllers Become AI Pioneers." *Science*, June 2, 1989.

"Fast, Cheap, and Out of Control." *Science*, May 25, 1990.

Wasser, Alan. "Power Tower." *Ad Astra*, October 1990.

Wickelgren, Ingrid. "Bone Loss and the Three Bears: A Circulating Secret of Skeletal Stability." *Science News*, December 24 & 31, 1988.

Will, Clifford. "The Binary Pulsar." *Mercury*, November-December 1987.

Zubrin, Robert M. "The Key to Mars, Titan and Beyond? Nuclear Rockets Using Indigenous Propellants." *Planetary Report*, May-June 1990.

Other Sources

"Aeritalia for Tethered Space Systems: Evolution of a New Technology for Space Applications." Booklet, Aeritalia Space Systems Group. September 1987.

"Applications of Tethers in Space." Proceedings of a workshop sponsored by the Italian National Space Plan, CNR and NASA. Venice, Italy: October 15-17, 1985. NASA Conference Publication 2422.

Cesarone, Robert J., Andrey B. Sergeyevsky, and Stuart J. Kerridge. "Prospects for the Voyager Extra-Planetary and Interstellar Mission." *AAS/AIAA Astrodynamics Specialist Conference.* Lake Placid, N.Y., August 22-25, 1983. San Diego: AAS Publications Office Paper no. 83-308.

"Columbus 500 Space Sail Cup Solar Sail Spacecraft." Technical Proposal (Volume 1). The Johns Hopkins University. November 1989.

Forward, Robert L. "Exotic Power and Propulsion Concepts." Keynote speech presented at Vision-21, Space Travel for the Next Millennium, NASA Lewis Research Center, Cleveland, Ohio, April 3-4, 1990.

Friends of the Goddard Library. *The Goddard Biblio Log*, First Supplemental Number, November 11, 1972. (Courtesy of the Goddard Library/Clark University Special Collections.)

Goddard, Robert H. "The Last Migration." Unpublished manuscript, January 14, 1918. (Courtesy of the Goddard Library/Clark University Special Collections.)

Holt, Alan C. "Spaceport Operations for Deep Space Missions." Paper presented at Vision-21, Space Travel for the Next Millennium, NASA Lewis Research Center, Cleveland, Ohio, April 3-4, 1990.

"An Investigation of the Needs and the Design of an Orbiting Space Station with Growth Capabilities." Final report on contract no. NASW 2776. Washington, D.C.: NASA-JSC, NASA-MSFC, January 1977.

Johnson, Richard D., and Charles Holbrow (eds.). "Space Settlements: A Design Study." Washington, D.C.: NASA Scientific and Technical Information Office, 1977.

NASA Office of Exploration:

"Beyond Earth's Boundaries." Annual Report to the Administrator. Washington, D.C.: NASA, 1988.

"Exploration Studies Technical Report, Volume 1: Technical Summary." Washington, D.C.: NASA, December 1988.

"Exploration Studies Technical Report, Volume 2: Study Approach and Results." Washington, D.C.: NASA, December 1988.

NASA Office of Space Flight. "Tethers in Space Handbook" (2d ed.). Washington, D.C.: NASA, May 1989.

Ride, Sally K., "Leadership and America's Future in Space." A Report to the Administrator. Washington, D.C.: NASA, August 1987.

Stern, Martin O. "Advanced Propulsion for LEO-Moon Transport." Progress report. California Space Institute, La Jolla: June 1988.

Thierschmann, M. "Comparison of Super-High-Energy-Propulsion-Systems Based on Metallic Hydrogen Propellant for ES to LEO Space Transportation." Vision-21, Space Travel for the Next Millennium, NASA Lewis Research Center, Cleveland, Ohio, April 3-4, 1990.

"Viking: Fact Sheet." Pasadena, Calif.: NASA, Jet Propulsion Laboratory. No date.

"Voyager: Mission Summary." Pasadena, Calif.: NASA, Jet Propulsion Laboratory. No date.

"Voyager: Neptune Science Summary." Pasadena, Calif.: NASA, Jet Propulsion Laboratory. No date.

Weidman, Deene J., William M. Cirillo, and Charles P. Llewellyn. "Study of the Use of the Space Station to Accommodate Lunar Base Missions." Lunar Bases and Space Activities in the 21st Century Symposium. NASA, et al.: April 5-7, 1988.

Zubrin, R., and D. Andrews. "Magnetic Sails and Interplanetary Travel." Monterey, Calif.: AIAA/ASME/SAE/ASEE, 25th Joint Propulsion Conference, July 10-12, 1989.

INDEX

ACKNOWLEDGMENTS

The editors wish to thank Buzz Aldrin, Laguna Beach, Calif.; Dana Andrews, Boeing Aerospace Co., Kent, Wash.; Colin Angle, Massachusetts Institute of Technology, Cambridge, Mass.; Dana Backman, NASA Ames Research Center, Moffett Field, Calif.; Ivan Bekey, National Space Council, Washington, D.C.; Andrew Cutler, Tucson, Ariz.; Leonard David, Space Data Resources and Information, Washington, D.C.; Gerald Feinberg, Columbia University, New York, N.Y.; Robert L. Forward, Malibu, Calif.; Peter E. Glaser, Arthur D. Little, Inc., Cambridge, Mass.; Goddard Library/Clark University Special Collections, Clark University, Worcester, Mass.; Bettie Greber, Space Studies Institute, Princeton, N.J.; Edward Krupp and Robin Rector Krupp, Eagle Rock, Calif.; Otto W. Lazareth, Department of Nuclear Energy, Upton, N.Y.; Christopher P. McKay, NASA Ames Research Center, Moffett Field, Calif.; Mary Fae McKay, Lyndon B. Johnson Space Center, Houston, Tex.; George Maise, Brookhaven National Laboratory, Upton, N.Y.; Marc G. Millis, Lewis Research Center, Cleveland, Ohio; Jacques-Clair Noëns, Observatoire du Pic-du-Midi, France; Bernard Oliver, Los Altos Hills, Calif.; Humphrey W. Price, California Institute of Technology, Pasadena, Calif.; Guillermo Trotti, Bell & Trotti, Houston, Tex.; Marie-Josée Vin, Observatoire de Haute Provence, France; Sam Wagner, General Dynamics Space Systems Division, San Diego, Calif.; Robert M. Zubrin, Martin-Marietta Astronautics Group, Denver, Colo.

PICTURE CREDITS

The sources for the illustrations that appear in this book are listed below. Credits from left to right are separated by semicolons, from top to bottom by dashes.

Cover: Art by Joe Bergeron. 2, 3: Art by Gaylord Welker. 8, 9: NASA, Photo No. 84-HC-264. 10: Initial cap, detail from pages 8, 9. 12, 13: Smithsonian Institution, Neg. No. 72-7456; Smithsonian Institution, Neg. No. 75-16236; Christopher Springmann. 14, 15: MAN, St. Germain-en-Laye, Photo R.M.N.; ©Robin Scagell/Science Photo Library/Photo Researchers; Robin Rector Krupp; courtesy Trustees of the British Museum, London. Background art by Stephen R. Wagner. 16, 17: Larry Robbins; Robin Rector Krupp; Michael Zeilik; Robin Rector Krupp. Background art by Stephen R. Wagner. 18, 19: Art by Fred Devita, based on sail design by Johns Hopkins University Applied Physics Laboratory. 20: Line art by Fred Holz, based on sail design by World Space Foundation—line art by Fred Holz, based on sail design by Cambridge Consultants, Ltd. 21: Line art by Fred Holz, based on sail design by Johns Hopkins University Applied Physics Laboratory—line art by Fred Holz, based on sail design by Massachusetts Institute of Technology. 22, 23: Art by Fred Devita. 25: NASA. 26, 27: NASA, Marshall Space Flight Center; TRW; art by Pat Rawlings/NASA. 28: Art by Pat Rawlings, courtesy Science Applications International Corporation. 32, 33: NASA, Photo No. 89-HC-319; NASA. 34, 35: NASA/JPL, Photo No. P-17430. 37: NASA/JPL Art, Photo No. P-35177. 38, 39: Art by Fred Holz. 42-51: Art by Joe Bergeron. Inset diagrams by Fred Holz. 52, 53: NASA/JPL. 54: Initial cap, detail from pages 52, 53. 57: Shel Hershorn/Black Star; UPI/Bettmann. 58: Art by Paul Hudson. 59: Art by Robert McCall. 61: Frank White. 63: Art by Denise Watt, courtesy John Dossey and Guillermo Trotti. 64: NASA Art, Photo No. 77-HC-432. 65: NASA Art, Photo No. 75-HC-272. 66: NASA Art, Photo No. 77-HC-405. 69: U.S. Geological Survey, Flagstaff, Ariz. 70, 71: NASA/JPL, Photo No. 211-5359A; NASA/JPL, Photo No. P-20942—NASA/JPL, Photo No. P-17698. 72: Art by Michael Carroll. 74, 75: Terry Smith/PEOPLE WEEKLY; Roger Foley; Michael Freeman, London. 77: NASA/JPL Art, Photo No. P-34311. 79: Art by Yvonne Gensurowsky—Linda McConnell/Rocky Mountain News; Boeing Aerospace & Electronics. 81: Art by Ron Miller. 84-91: Art by Stephen R. Wagner. 92, 93: Royal Observatory, Edinburgh, and Anglo-Australian Telescope Board. 94: Initial cap, detail from pages 92, 93. 96, 97: NASA, Photo No. 73-HC-61; NASA/JPL, Photo No. P-24653. 100, 101: Talbot Lovering, courtesy MIT Mobile Robotics Laboratory. 105-109: Art by Rob Wood of Stansbury, Ronsaville, Wood, Inc. 114-119: Line art by Fred Holz. 122, 123: Art by Matt McMullen. 126-133: Art by Fred Devita.

Time-Life Books Inc.
is a wholly owned subsidiary of
THE TIME INC. BOOK COMPANY

TIME-LIFE BOOKS

PRESIDENT: Mary N. Davis

Managing Editor: Thomas H. Flaherty
Director of Editorial Resources:
Elise D. Ritter-Clough
Director of Photography and Research:
John Conrad Weiser
Editorial Board: Dale M. Brown, Roberta Conlan,
Laura Foreman, Lee Hassig, Jim Hicks, Blaine
Marshall, Rita Thievon Mullin, Henry Woodhead
Assistant Director of Editorial Resources/
Training Manager: Norma E. Shaw

PUBLISHER: Robert H. Smith

Associate Publisher: Trevor Lunn
Editorial Director: Donia Steele
Marketing Director: Regina Hall
Production Manager: Marlene Zack
Supervisor of Quality Control: James King

Editorial Operations
Production: Celia Beattie
Library: Louise D. Forstall
Computer Composition: Deborah G. Tait
(Manager), Monika D. Thayer, Janet Barnes
Syring, Lillian Daniels
Interactive Media Specialist: Patti H. Cass

Correspondents: Elisabeth Kraemer-Singh (Bonn),
Christine Hinze (London), Christina Lieberman
(New York), Maria Vincenza Aloisi (Paris), Ann
Natanson (Rome). Valuable assistance was also
provided by Elizabeth Brown (New York), Judy
Aspinall (London).

VOYAGE THROUGH THE UNIVERSE

SERIES EDITOR: Roberta Conlan
Series Administrator: Norma E. Shaw

Editorial Staff for *Starbound*
Art Director: Dale Pollekoff
Picture Editor: Kristin Baker Hanneman
Text Editors: Stephen Hyslop,
Robert M. S. Somerville
Associate Editor/Research: Karin Kinney
Assistant Editors/Research: Dan Kulpinski,
Quentin Story
Writers: Darcie Conner Johnston, Barbara Mallen
Assistant Art Director: Barbara M. Sheppard
Editorial Assistant: Katie Mahaffey
Copy Coordinator: Juli Duncan
Picture Coordinators: Barry Anthony, Betty H.
Weatherley
Special Contributors: Sarah Brash, Deborah Byrd,
George Constable, James Dawson, Jeff Kanipe,
Michael Lemonick, Gina Maranto, Chuck Smith,
Elizabeth Ward, Mark Washburn (text); Vilasini
Balakrishnan, Craig Chapin, Mark Cheater, Nancy
Connors, Jocelyn G. Lindsay, Ted Loos, Cheryl
Pellerin, Eugenia Scharf, Jacqueline Shaffer
(research); Barbara L. Klein (index).

CONSULTANTS

FRANK DRAKE is an astronomer who teaches at
the University of California at Santa Cruz. He is a
founder and the president of the SETI (Search for
Extraterrestrial Intelligence) Institute.

HAROLD G. FOX, an engineer, is the program manager of the Solar Sails Project and supervisor of the
Environmental Test Facility at Johns Hopkins University.

VICTORIA GARSHNEK specializes in space medicine at NASA Headquarters in Washington, D.C. She
also teaches at the Space Policy Institute at George
Washington University.

RICHARD E. GERTSCH, a mining engineer, is a
member of the Center for Space Mining, Colorado
School of Mines. He has been a space mining con-
sultant for NASA and has taught courses in the
subject.

BRAND NORMAN GRIFFIN, an aerospace designer,
is currently a project manager at Boeing Aerospace
and Electronics in Huntsville, Alabama, studying
the future exploration of the Moon and Mars.

ALAN HOLT, a physicist and an engineer, is deputy
manager of the Payload Integration Office, NASA
Headquarters. He manages the integration of payloads on the space station *Freedom*.

TED MAXWELL is chairman of the Center for Earth
and Planetary Studies, National Air and Space Museum, Smithsonian Institution. He is also director
of the Smithsonian's Regional Planetary Image Facility.

PAUL A. PENZO is a staff scientist at the Jet Propulsion Laboratory of the California Institute of
Technology. He has performed mission design studies for the Apollo, Voyager, and Galileo projects, and
more recently for tethers in space.

ERIC SHELDON, who teaches physics, relativity
theory, and cosmology at the University of Lowell in
Massachusetts, has designed a computer program
that shows what the starscape would look like from
a spacecraft moving at near light-speed.

BRIAN TILLOTSON works in the Advanced Civil
Space organization of Boeing Aerospace and Electronics in Huntsville, Alabama. A specialist in artificial intelligence, robotics, and advanced propulsion, he is also the inventor of the multistage tether
concept.

**Library of Congress Cataloging in
Publication Data**
Starbound / by the editors of Time-Life Books.
p.   cm.   (Voyage through the universe).
Bibliography: p.
Includes index.
ISBN 0-8094-6941-3.
ISBN 0-8094-6942-1 (lib. bdg.).
1. Outer space—Exploration.
I. Time-Life Books. II. Series.
QB500.262.S73   1991
919.9'04—dc20   90-47598 CIP

For information on and a full description of any
of the Time-Life Books series, please call 1-800-
621-7026 or write:
Reader Information
Time-Life Customer Service
P.O. Box C-32068
Richmond, Virginia 23261-2068

Earth: diameter 7,926 miles

Neptune: diameter 30,775 miles

Uranus: diameter 31,763 miles

Red supergiant: diameter 400 million miles

Solar System: diameter 7.5 billion miles

Globular cluster: diameter 2×10^{14} miles

Milky Way: diameter 100,000 light-years

Local Group of galaxies:
6 million light-years across

Largest double radio source:
length 17 million light-years